THE OCEAN

OUR FUTURE

**The Report of the Independent
World Commission on the Oceans**

Chaired by Mário Soares

CAMBRIDGE

PUBLISHED BY THE PRESS SYNDICATE OF THE UNIVERSITY OF CAMBRIDGE
The Pitt Building, Trumpington Street, Cambridge CB2 1RP, United Kingdom

CAMBRIDGE UNIVERSITY PRESS
The Edinburgh Building, Cambridge CB2 2RU, UK. http://www.cup.cam.ac.uk
40 West 20th Street, New York, NY 10011-4211, USA. http://www.cup.org
10 Stamford Road, Oakleigh, Melbourne 3166, Australia

First published 1998

Printed in the United Kingdom at the University Press, Cambridge

Typset in Stone Serif 9.5/12pt

A catalogue record for this book is available from the British Library

ISBN 0 521   64286   8   hardback
ISBN 0 521   64465   8   paperback

Cover Photo Captions and Credits:

Cover: [Main] Reflections and sun rays on seascape, France
(Gilles Corniere); [Inset] Jellyfish *Cassiopeia sp.* (Norbert Wu).

Back cover:
[Main] Wave off the coast of Hawaii (FOTO/UNEP)
[Inset] Earth from space showing South-East Asia,
Australia and Pacific Ocean (DRA)

All photos copyright: Still Pictures

THE OCEAN OUR FUTURE

**The Report of the Independent
World Commission on the Oceans**

Chaired by Mário Soares

CAMBRIDGE
UNIVERSITY PRESS

INDEPENDENT WORLD COMMISSION ON THE OCEANS

This Report expresses the collective views of the members of the Commission. This does not necessarily imply that all members subscribe to the details of each of its formulations.

CONTENTS

ACKNOWLEDGMENT

The Commission is grateful for the assistance it has received, in the preparation of this Report, from numerous persons and organizations. Acknowledgment is specially due for financial support from a variety of sources from twelve countries, located in different regions of the world, as well as from several international organizations. Without this assistance and support, which is described in detail in Annex E, the Commission would not have been able to perform its task.

Preface

Although I am not an expert on the oceans – far from it – the sea has always interested and motivated me. I was born and have nearly always lived near the sea, in a city – Lisbon – which lies on the estuary of a great river that flows into the Atlantic. Here, at all times of day and in all weathers, the sea and the maritime environment are a constant, pervasive and complex presence. This presence has always affected me deeply, as a source of energy and inspiration, a vehicle of myth, a spur to action, an encouragement and invitation to reflection. To be near the sea is so vital to me that when I was exiled in Paris I would often travel 200 kilometres and more just to see, to hear and to breathe the sea – those bleak, grey northern waters so different from the sea of my own country – a deep blue or clear viridian depending on the hour of day.

I come from a country of mariners whose history is steeped in the heroism of the discoveries which unified the world, linking east and west; a country of successive generations of traders, explorers and missionaries, who adventured in all directions across an uncharted sea. The magnificent accounts they left of their voyages around the world fascinated contemporary Europe, and even today give cause for admiration. As a country of poets, Portugal has celebrated the sea in different ways and in different lights, in all the terror, hope and enticement it inspires, and also in the tragedy, suffering and anguish with which the maritime history of the country is replete. 'Salt-laden sea, how much of your salt is the tears of Portugal', wondered Fernando Pessoa, a poet at once esoteric and universal, visionary, lucid – and Portuguese.

I belong, body and soul, to an old nation with a strong cultural identity which, for nearly nine centuries, wedged between Spain and the Atlantic, has maintained the same frontiers. This, no doubt, is why I have always seen the ocean as the cause of, and the explanation for, the uniqueness of the Portuguese spirit. But, more importantly, I have always seen the ocean as a realm of freedom, a prime medium of human contact – and therefore of dialogue, solidarity and coexistence between different cultures and civilizations, a coming-and-going of influences measured by the rhythm of arrivals and departures.

In politics, I served my apprenticeship in the long and hard battle against the oppression of the ultra-conservative, colonialist dictatorship, which controlled the people of Portugal and the Portuguese-speaking countries of Africa for almost 50 years. It was an uneven battle. For this reason, I have always fought much more for freedom – in defence of human rights and the great issues of the day – than for power.

Preface

I have always seen the oceans as a school for democracy on a global – but also regional and even national – scale; a school for co-operation, understanding and common security. Furthermore, contrary to what occurs with terrestrial resources, which can be individually possessed and appropriated in forms developed and consecrated over the centuries, marine resources are by their own nature common, and are generally considered as such. Even in the exclusive economic zones over which coastal states exercise sole jurisdiction, the decisive importance of new marine technology – and the scientific sophistication which its full and effective application requires – entails not only the notion of responsible sovereignty but also that of sharing and cooperation with other states, namely the less-developed ones, which do not yet have the know-how and, above all, the necessary means. We need therefore, to forge a new ethico-political relationship between humanity and the oceans, a relationship with a political and juridical basis which creates an atmosphere of sharing and solidarity and which provides for a new universalism centred on the knowledge of the oceans; a relationship capable of unifying the citizens of the world under one banner, a common, unique and irreplaceable asset: the sea which all the continents share, and which to a certain extent equalizes them.

As a former political leader in Portugal, I have valued the importance of the oceans and the urgent need for Portugal's return to the sea ever since the 'Revolution of the Carnations' which brought the country its freedom. Now that the cycle of colonialist expansionism is over, this return to the sea assumes, in my view, an importance which is practically a vocation. As a European country actively committed to the construction of the new Europe, Portugal has an irrevocable duty to uphold the importance of the oceans for the European Union. And, as a European, I have sought via different non-governmental organizations to heighten the awareness of decision-makers at various levels, and of the public in general, of the advantages and timeliness of creating a European Agency for the Oceans. An Agency for the coordination and optimization of the efforts of the countries of the Union in the race for the oceans which the coming century will inevitably see.

This race, in my view, is not one which can simply be left to the capacity of enterprise and the greater opportunities enjoyed by countries which are technologically more advanced. It should be based on just and effective international legislation, on equity, solidarity and sharing. The United Nations Convention on the Law of the Sea, already ratified by around 130 countries, is a good point of departure and deserves to be followed through. As a citizen

Preface

concerned with the direction the world is taking, a direction which promises so much yet at the same time leaves so much uncertainty, I have participated in countless regional and global meetings and conferences with the objective of alerting public opinion, policy-makers and economic agents to the crucial importance of the oceans in forthcoming decades. The oceans represent a source of enormous potential for humanity, not only in material but also in spiritual terms, yet they are increasingly falling prey to intolerable aggression with grave consequences for the survival of the human race.

We urgently need therefore, to learn together how to preserve the oceans with humility and with effectiveness.

The regeneration capacity of the oceans is enormous. But it is not, contrary to what many people imagine, unlimited. Many of its most significant living resources are being over-exploited, with a reprehensible lack of foresight. Non-living resources are equally at risk and are often used up with scant attention to environmental regulations and in ignorance of the ecosystems which such form of exploitation necessarily threatens.

In today's world, the oceans represent a new frontier which we must open up with care if humanity as a whole, and not just the richer and more developed countries, is to benefit. We must remember too that deep in the ocean lies one of the main archives of human history that must be interpreted and preserved.

It was doubtless for some of the above reasons that the former Secretary-General of the United Nations, Boutros Boutros-Ghali, and the current Director-General of UNESCO, Federico Mayor Zaragoza, encouraged me to form the Independent World Commission on the Oceans, which was finally launched at the United Nations University in Tokyo in December 1995. Kofi Annan, the present Secretary-General of the UN, has expressed his interest in the collective work carried out since then. This has been, for us, an important gesture of support.

What did all this involve, then? The creation of a *global* and *independent* Commission – independent relative to governments and international organizations – whose objective is to foster critical reflection on the many aspects of the current situation of the oceans from an integrated and multidisciplinary perspective. A Commission which – like the one Willy Brandt created on the North–South dialogue – should be capable of making a critical evaluation of the state and trends of ocean affairs, setting forth clear recommendations. It should serve as an alarm and speak a language which everyone can understand; it should appeal for true international cooperation in the preservation of the oceans as an integrated ecosystem whose

Preface

unique way of functioning we must respect. And it is an initiative which is all the more timely in that the United Nations has designated 1998 as the International Year of the Ocean, following a proposal initiated by Portugal. World attention is therefore, focused on the Oceans.

We need to ensure that management of the oceans is rational, just and responsible *vis-à-vis* future generations. We need to assess the enormous potential of science and the new technologies for the sustained use of marine resources. We need to find practical ways of sharing this potential with the less-developed countries in a spirit of cooperation and democracy – ways which can create an integrated and consensual governance of the oceans under the aegis of the United Nations.

The end of the cold war radically changed the geo-strategic orientation of the major maritime powers. The risk of world conflict has been eliminated: but the same cannot be said of regional conflicts, which threaten to escalate. Similarly, the increase of piracy, arms and drugs trafficking, and illegal and flagrant use of the sea as a dumping ground for toxic and nuclear waste, are causes for concern. However, the declassification by the United States of major ocean-ographic databases and the use of its naval infrastructure for the monitoring of the oceans are welcome initiatives. In this context, we believe that the European Union should enhance the civilian role of the naval forces of its Member States, thus encouraging the peaceful use of the oceans.

As we approach the new millennium, it is more evident than ever before that the oceans are a common asset of humanity as a whole. The oceans are a privileged space for the strengthening of relationships between states: relationships forged on a spirit of cooperation, understanding and solidarity. With an economic approach prevailing in these days of harsh competition, the important capital that the oceans represent to humanity is often overlooked, particularly their non-material aspects. This capital has no price, no replacement and no exchange value. We must preserve it for the benefit of present and future generations.

The principle of sustainable development adopted at the Earth Summit of Rio de Janeiro in 1992 has yet to be applied internationally in an effective manner. Six years after Rio, the results are insufficient and even disappointing, as the recent Kyoto Conference made clear. With a few honourable exceptions, governments, parliaments and international organizations have proved incapable of implementing the necessary measures for the recommendations, the declarations of intention and the principles proclaimed to be generally observed.

Preface

This is not a question of drawing up international legislation. It already exists. It is a question, rather, of the practical application of this legislation and of penalizing those who infringe the law.

Despite the globalization of the economy – with all that this entails, both negative and positive – we must look further than free trade and competition. What really matters are the human beings – men, women and children – who should be the end recipients of progress but who are the victims, rather than the beneficiaries, of this progress: unemployment is growing, poverty is increasing and inequality, in every continent on earth, is becoming explosive.

In open and democratic societies where information circulates and freedom of expression is a right – and this is a huge step forward from the closed, totalitarian societies of the recent past – we have to appeal to the citizen and to civil society in instances where policy-makers, business and international organizations prove incapable of finding solutions to the problems facing today's world. Science and technology have the means to solve these problems: if the political will exists.

'The Ocean ... Our Future', the Report presented here under the collective responsibility of the Independent World Commission on the Oceans, aims to be a practical and realistic appeal to the conscience of the citizens of the world. History's judgement of the present generation may be extremely harsh where the preservation of the oceans is concerned. It is vital therefore, that we reverse the dominant tendencies. We appeal directly to non-governmental organizations, policy-makers and businesses on a national and international level. The byword of the coming century will be solidarity – I am quite convinced of this. The effective and lasting preservation of the oceans is a task which needs solidarity; it directly affects the future and the generations to come. And it cannot be postponed for there is no time to lose.

Summary

The oceans have traditionally been taken for granted as a source of wealth, opportunity and abundance. The vastness of ocean space that fuelled our inspiration and curiosity, suggested that there could be few if any limits to its use or abuse. Our growing understanding of the oceans has fundamentally changed this perception. It has led to a growing appreciation not only of the importance of the oceans to social and economic progress but also of their vulnerability. We now know that abundance is giving way to scarcity, in some cases at an alarming rate, and to conflicts arising from their use.

Life on our planet is dependent upon the oceans. They provide us with food, energy and water and they sustain the livelihoods of hundreds of millions of people. They are the main highway for international trade as well as the main stabilizer of the world's climate. However, in the space of only a few decades the oceans have become the setting for an expanding list of problems. Territorial disputes that threaten peace and security, global climate change, overfishing, indiscriminate trawling, habitat destruction, species extinction, pollution, illegal trafficking, congested shipping lanes, clandestine movement of persons, piracy, terrorism, and the disruption of coastal communities are among the problems that today form an integral part of the unfolding drama of the oceans. At the same time, the oceans are revealing to us great potentials and opportunities.

The challenge posed by the oceans is one of truly historical dimensions, since the extent to which it is met will have a major bearing on the well-being not only of almost everyone alive today but also of future generations.

THE COMMISSION

The Independent World Commission on the Oceans was launched in December 1995. It reviewed the existing situation and identified future directions. It presents its recommendations in 1998, the International Year of the Ocean, to the international community as a whole and to the UN General Assembly. Conscious of the size of the challenge and complexity of the problems to be addressed, the Commission has adopted an approach that is deliberately selective in scope. In looking to the future, the Commission has defined its task as one of identifying directions with a strategic significance or which could guide future action and debate both within and outside the intergovernmental system.

Summary

The findings in this report provide cause for hope as well as concern. The Commission holds the view that the oceans are under sustained pressure and that it is no exaggeration to refer to a crisis in the oceans. This crisis cannot be seen in isolation from the many problems that affect both land and air. Indeed, together they form part of the *problematique* of our biosphere in which issues at sea are connected to those on land through rivers, the atmosphere and the coastal zone. All form part of a larger picture that links unsustainable resource use to the well-being of future generations and, ultimately, to prospects for human survival.

The problems in the oceans are multi-faceted. They have many dimensions, including moral and ethical ones. These find tangible expression in the inequalities of opportunity existing between rich and poor nations and in the absence of mechanisms to ensure that all nations and peoples benefit equitably from the uses of the oceans and the exploitation of their resources. They are also manifest in the failure to safeguard the interests of future generations, not only of those who will be born into poverty and underdevelopment but also of the offspring of today's affluent and privileged minority. There are issues of fairness that must be addressed in relation to the oceans. To do so calls for solidarity with present and future generations.

The fact that the world community has so far failed to recognize the seriousness of the deterioration that has taken place is regarded by the Commission as a major obstacle to change.

FUTURE DIRECTIONS

'Conscious that the problems of ocean space are closely interrelated and need to be considered as a whole' as well as in relation to land-based activities, the Commission has chosen to highlight several issues where major adjustments and innovations will be required if obstacles to change are to be addressed effectively. These issues are grouped under six main headings, which establish the structure for the report:

- promoting peace and security in the oceans;
- the quest for equity in the oceans;
- ocean science and technology;
- valuing the oceans;
- our oceans: public awareness and participation;
- towards effective ocean governance.

Summary

Promoting peace and security in the oceans

In the changed post-Cold-War setting, there are numerous threats in the oceans, other than military ones, affecting the security of nations and peoples – such as pollution of the marine environment, unsustainable use of ocean resources, illicit trafficking, clandestine movement of persons, piracy, terrorism, congested shipping lanes – and it is in the interest of all to devise forms of cooperation to deal with them more effectively. Given the provisions of the 1982 UN Convention on the Law of the Sea for extended coastal state jurisdiction, a major challenge to efforts to advance peace and security is to be found outside national jurisdiction, that is on the 'high seas'. In order to better meet this challenge, the Commission recommends that:

• The 'high seas' be treated as a public trust to be used and managed in the interests of present and future generations.

• The role of navies and, where appropriate, other maritime security forces be reoriented, in conformity with present international law, to enable them to enforce legislation concerning non-military threats that affect security in the oceans, including their ecological aspects. Navies could also play a growing role in sharing the information and capabilities required to safeguard environmental security.

Peace and security in the oceans can be advanced by other measures including: a greater commitment to the peaceful resolution of territorial disputes, for instance, through 'freezing' of contentious claims and the creation of 'joint management zones'; more determined efforts to establish effective nuclear-free zones; and an updating of the law of naval warfare. The Commission is also of the opinion that progress would be served by the preparation of a Report on Peace and Security in the Oceans in the Twenty-first Century that would extend to the oceans the thinking contained in the 1992 UN Secretary-General's Report, *An Agenda for Peace*.

The quest for equity in the oceans

Greater equity in the oceans would contribute to reducing poverty and underdevelopment in general. Notwithstanding the fact that the oceans are under sustained pressure, we must recognize that they continue to offer an immense source of wealth. Although the economic value of the sea defies easy measurement, it is clear that

Summary

the oceans already contribute very significantly to social and economic development and that their future contribution could be greater than we now imagine.

At present, the benefits from the exploitation of marine resources are shared inequitably between the world's nations, and inequalities will continue to exist until such time as mechanisms are established that provide more effectively for benefit sharing. With this in mind, the Commission recommends that:

● The oceans be regarded as a common resource so as to make it possible for nations and people to share more equitably in the benefits of resource exploitation and to accelerate the social and economic development of disadvantaged nations.

● Initiatives be considered, through the Global Environment Facility or other means, for less-developed coastal states aimed at building the capacity required for them to take effective advantage, in a sustainable manner, of their exclusive rights to the use of the resources in areas under their jurisdiction and to meet their concomitant obligations under the Law of the Sea Convention.

● Special measures be adopted to protect vulnerable groups, especially indigenous peoples and local communities dependent for their livelihoods on subsistence fishing, and that commercial fishing agreements take more fully into account the needs of those communities through provisions for indemnification and local capacity building.

Equity in the oceans should be enhanced by initiatives in other areas, including: the establishment of regional systems for sustainable development and related marine science and technology; action-orientated studies on the economic, environmental and legal implications of the genetic resources associated with marine hydrothermal vents beyond national jurisdiction; new initiatives in resource mobilization, based on the use of the oceans as a source of development finance; and the enforcement by governments of international rules governing the security of marine traffic, the operational and environmental safety of ships and the working conditions of seafarers.

Summary

Ocean science and technology

The application of modern technology to the oceans, when ill-considered, is linked to their deterioration and overexploitation. At the same time, modern technology is the most powerful force for translating potentials into reality and for satisfying ever-growing basic needs. It is also central to our efforts to acquire the scientific knowledge to better understand the oceans and the relationships between the seas and human activity. With this in mind, the Commission recommends that:

• Science and technology be directed to a greater extent towards translating the potential of the oceans into the satisfaction of basic needs.

• Systematic efforts be made to subject technologies for the exploration and exploitation of marine resources to prior assessment of their environmental and social impacts in order to minimize negative effects on the oceans and coastal zones.

• Greater emphasis be given to initiatives aimed at enlarging the access of developing countries to scientific information and technologies, especially in sub-regional and regional settings and in ways that take advantage of modern systems of information dissemination and exchange.

Opportunities must also be created that enable developing countries to participate more fully in scientific research and exploration, especially in mega-science projects, such as the Global Ocean Observation System, deep ocean drilling and the study of the genetic resources of the deep seabed.

Valuing the oceans

In the Commission's view, past approaches to the economics of the oceans have been short-sighted. There has been a consistent underestimation of the value of the oceans and of the ecological services they provide. As a result, the uses of the oceans have failed to take into account external costs, which has contributed to unsustainable levels of exploitation of resources and to the rapid deterioration of the marine environment.

Summary

A future approach must embody an unequivocal commitment to safeguarding the health of the oceans and the productivity of ocean ecosystems. It should be guided by the 'precautionary principle' and the readiness to recognize the intrinsic ecological value of the oceans, to internalize the full range of external effects in decision-making on the oceans, and to withdraw subsidies that contribute to the deterioration of the oceans and to unsustainable patterns of resource exploitation. Future approaches must also recognize the importance of establishing appropriate management regimes through international agreements, and should be supported by the refinement of methodologies for the systematic valuation of the oceans and their resources. To this end, the Commission recommends that:

● Efforts be made to ensure, by applying the 'user-pays' and 'polluter-pays' principles, that the users of ocean resources and polluters of marine ecosystems bear the true costs of their actions.

● Wherever appropriate, incentives – environmental taxes, user charges, etc. – be introduced to encourage the sustainable use of the oceans, and to eliminate subsidies that encourage waste and overuse of ocean resources.

● Management regimes embodying the precautionary principle be established at the appropriate geographical level. These regimes should also recognize the importance of a multi-sectoral and multi-disciplinary approach and of on-shore/off-shore linkages.

Our oceans: public awareness and participation

To raise world consciousness of the oceans has been an important part of the mandate of the Commission. Efforts to increase public awareness depend on more transparency in ocean affairs, giving unambiguous expression to the public's 'right to know' what is happening to, on and beneath the surface of, the sea. Progress in this area will, in the view of the Commission, be contingent upon the creation of arrangements which ensure that information and knowledge are more freely available for public discussion on the future of the oceans. It is part of our intergenerational responsibility to transmit this knowledge to children and young people, so as to enable them to appreciate the vital importance of the oceans, the values they represent and the risks they face.

Summary

Beyond the basic 'right to know', a number of channels also exists for civil society to exercise the 'right to be heard' – as well as the 'right to complain' – in international ocean affairs. These channels should be activated and strengthened, with a view to defusing conflicts and facilitating the interaction of non-state entities with the intergovernmental system. In this context, at the global level, consideration should be given to the appointment of an independent Ocean Guardian, with a mandate to take up grievances concerning alleged non-compliance with international marine agreements, or misuse of the oceans and their resources.

Furthermore, despite some promising initiatives, the reality of public participation in ocean governance today remains a far cry from the ambitious expectations created by the 1992 UN Conference on Environment and Development (UNCED) and its Agenda 21. One reason for this disappointing record is the lack of independent representative institutions qualified to speak out on behalf of universal values – on behalf of civil society, future generations, the global environment, or of other public interests. A more informed and active civil society, with significantly expanded opportunities to participate in ocean affairs, is a precondition for a more democratic, responsive and coherent system of ocean governance.

Towards effective ocean governance

The overarching challenge that must be addressed is one of developing systems of ocean governance that promote peace and security, equity and sustainable development. In the view of the Commission, the foundation for more effective ocean governance must be the Law of the Sea Convention and its 'implementing agreements'. The Convention, which finally entered into force in 1994, ranks as one of the major achievements of the international community in recent decades. The Commission also acknowledges the impetus given to the desired process of change by Agenda 21 and other global programmes and agreements – such as on climate change, biodiversity, and land-based sources of marine pollution – as well as by agreements concluded at the regional and sub-regional level.

However, ocean governance must be based on more purposeful and responsive policies and programmes for the coastal zone, the critical interface between people and the sea. It must also provide for action at the local, national, regional and global levels, especially in the management of human activities that, directly and indirectly, have long-lasting negative impacts on the health of the oceans.

Summary

Cooperation organized at the regional level carries particular promise for the more sustainable management of marine resources and of the marine environment.

The Commission is of the opinion that efforts to build a more effective system of ocean governance must start with the implementation of the Convention as well as the many other legal instruments existing in relation to the sea. A stronger 'political will' has to be displayed to ensure full compliance with existing ocean law and the adoption of effective measures of enforcement. In order to contribute to building a more effective system of ocean governance, the Commission recommends that :

• The discussion of ocean affairs within existing fora of the UN system be strengthened and supplemented by a comprehensive review of the mandates and programmes of all UN bodies and agencies competent in ocean affairs.

• The process of change and innovation within the intergovernmental system be facilitated by convening, at an early opportunity, a United Nations Conference on Ocean Affairs. This Conference would aim at placing the oceans prominently on international and national political agendas. It would take as its basis the Law of the Sea Convention, as well as other relevant international treaties and programmes, but would not be law making.

The recommendations set out above are essentially directed at states in recognition of the key role they must play in shaping a more effective system of ocean governance. They are presented, quite deliberately, with an eye to political feasibility and, together, they would reinforce the present system, making it more coherent, more responsive and more democratic.

However, this report also argues that the challenges posed by the oceans will not and cannot be adequately met unless *global civil society* is provided with significantly expanded opportunities to participate in ocean affairs. Change and innovation within the governmental and intergovernmental systems also call for initiatives outside these systems. This plea should not be regarded as a plea for the creation of institutions that would compete with governmental bodies or which would seek to duplicate their work. Rather, they should be interpreted as complementary measures that seek to make the system of ocean governance more responsive and more democratic. The Commission considers that it is essential to ensure independent

Summary

monitoring (to enhance *transparency*) and independent assessment (to enhance *accountability*) of ocean affairs. Accordingly, the Commission recommends:

● The establishment of a *World Ocean Affairs Observatory* in order to independently monitor the system of ocean governance and to exercise, on a continuous basis, an external watch on ocean affairs.

In the first instance, the Observatory would serve as a focal point for bringing together relevant information from various sources – official and unofficial, including intergovernmental, governmental and non-governmental institutions or networks. The information so obtained would be used by the Observatory to produce periodic 'state of the oceans' reports as well as *ad hoc* studies of urgent ocean issues. At the same time, the Observatory would serve as an interactive 'virtual' observation site for all ocean-related information on the World Wide Web, providing direct electronic links to all relevant (public and private) Internet sites.

Today, there are a number of precedents for successful performance of this monitoring role – in the fields of human rights, environment and disarmament – by non-governmental bodies such as Amnesty International, Greenpeace and SIPRI. These independent bodies play a crucial 'watchdog' role, and hence may serve as precedents for the work of the Observatory.

As a complementary measure, the convening of an *Independent World Ocean Forum* would allow public assessments by an independent assembly representing civil society and all stakeholders. It would further allow the actors to be held accountable for the use of ocean space and the management of its resources. The Forum would be outside of the intergovernmental structures previously discussed; it would have no decision-making powers; and it would operate as a 'recurrent event' (taking place every three or four years) rather than as a permanent institution. The Forum would draw on studies and other outputs generated by the Observatory, and on interactive electronic communications with open public participation during pre-sessional and intersessional periods.

Such initiatives would enable those with an interest in the oceans – and their many interactions with land-based activities, rivers, and coastal areas – to better articulate their concerns and express their hopes and aspirations. They would empower new voices to speak up for the oceans.

Today we celebrate the entry into force of the United Nations Convention on the Law of the Sea... . The dream of a comprehensive law for the oceans is an old one. Turning this dream into reality has been one of the greatest achievements of this century. It is one of the decisive contributions of our era. It will be one of our most enduring legacies... The achievement we celebrate can provide the international community with a new impetus, and a new opportunity. It can help humanity to realize the enormous potential of our common inheritance.

Boutros Boutros-Ghali, Secretary-General of the United Nations,
16 November 1994

This planet does not belong to the adults of today and should not be managed on the basis of short-term considerations of economic gain or political power. If the signatures of our children were needed to ratify decisions that affect their future, many of the destructive actions perpetrated today would certainly cease. Whatever we do, the ocean will survive in one way or another. What is more problematic is whether we shall preserve it in a state that ensures humanity's survival and well-being.

Federico Mayor, Director-General of UNESCO, 1998

Introduction

The oceans have traditionally provided a source of inspiration and myth, constituting a virtually unbounded domain that, from time immemorial, has nurtured fantasy, superstition and fear as well as curiosity, hopes and aspirations. The vastness of the oceans has filled human imagination with notions of eternity and responded to the deep desire for adventure. The same vastness has suggested an abundance of resources that exceeds human capacity for use and abuse. Unlike the land, where periods of scarcity have not been uncommon, the oceans have promised wealth and rewards that seem inexhaustible.

Our growing understanding of the oceans, made possible by rapid advances in science and technology, has fundamentally changed our perception of the oceans based on such innocence. It has changed with a speed that has few historical precedents and which challenges our capacity to devise appropriate strategies of adaptation.

In order to understand the profound change that has taken place in the way in which the oceans are perceived, it is necessary to look briefly at the past. For centuries, the use of the oceans and their resources was guided by a permissive regime based on the implicit assumption that it was possible to accommodate virtually all uses. Although there were problems in practice – notably in the form of piracy – that added to the risks imposed by nature, the assumption appeared a valid one. Fish and other species were generally plentiful, wastes disposed of at sea caused only local and temporary difficulties, navigation was unimpeded, and beaches remained largely unaffected by the uses of the sea.

This setting gave birth to the notion of the freedom of the seas predicated on the assumption that the exercise of the right to use the oceans was unlikely to infringe upon the same right held by others. A minor exception was made so as to provide coastal states with the opportunity to defend their territory. A territorial sea, with a width of three miles (the range of cannon shot in the seventeenth century), was the only area in which coastal states were empowered to exert their authority, although, even in the territorial sea, ships of other countries, including naval vessels, enjoyed unrestricted rights of navigation provided they conformed to accepted practices. Under these conditions of freedom and abundance, there was little need for law, and the law that existed was, for centuries, based on customary practices.

Introduction

SHIFT IN BASIC CONDITION OF THE OCEANS

The situation began to evolve this century, with the speed of change accelerating significantly in the past few decades. The main reason for this has been a shift in the basic condition of the oceans from one of apparent abundance to one of growing scarcity and from one of accommodation to one of conflict. A variety of factors has contributed to this shift.

First, the growth in the intensity of ocean use and in the number and magnitude of activities threatens to severely impinge on the carrying capacity of the oceans and on levels of sustainable use. Fishing grounds that were formerly very productive have, for example, become seriously depleted and some habitats have been irreversibly destroyed. Some coastal communities that have, for generations, been dependent upon the sea have not only lost their source of livelihood but also the meaning to their lives.

Secondly, the capacity of the oceans to accommodate increasing demands has been steadily eroded, and conflicts among competing uses have become commonplace. Experience reveals that there are inherent conflicts between the exploitation of the resources of the seabed and the preservation of the living resources of the water column above it. A few hours of indiscriminate trawling are, for example, sufficient to destroy a million years of coral growth and the assets required to support marine tourism or to maintain the livelihoods of traditional fishing communities.

Thirdly, the oceans can no longer be considered to be existing in isolation from the land. The health of the oceans has deteriorated mainly as a consequence of the pollution caused by land-based activity which is carried to the sea not only by rivers but also by the atmosphere. Of the many thousands of chemicals that are used for different purposes, most end up in the oceans and, overall, around 77% of marine pollution is estimated to have its origins on land, indicating that it is increasingly necessary to think in terms of systems that include both the oceans and river basins.

Fourthly, the oceans are threatened by the release or deliberate dumping of extremely hazardous wastes, such as pesticides, heavy metals and radioactive residues, by risks associated with the carriage of plutonium and dangerous chemicals, and by the possibility of accidents involving nuclear warheads or nuclear-powered vessels.

Fifthly, many negative developments are linked to the rapid growth in the world's population and, especially in the industrialized countries, to the even more rapid growth of economic output. When conflicts in the use of the oceans began to emerge early last century, the earth's population numbered around 1 billion. By the middle of this

Introduction

Sources of marine pollution

A very approximate estimate of the relative contribution of all potential pollutants from various human activities entering the sea could be summarized as follows:

Source	% Contribution	
Off-shore production	1	
Maritime transportation	12	
Dumping	10	
Sub-total ocean-based sources		23
Run-off and land-based discharges	44	
Atmosphere	33	
Sub-total land-based sources		77
Total all sources		100

These figures clearly demonstrate that marine pollution is derived mainly from land-based sources, directly or through the atmosphere. The impacts of these two types of sources are, however, very different. Atmospheric inputs to the seas are normally dilute and diffuse, while direct land-based inputs are often from point sources and can have long residence times in waters which are relatively enclosed by either geographic or hydrographic structures.

The relative contributions from each source are different in different sea areas, as these contributions depend on the degree of industrialization, the density of population, the extent of off-shore activities and other factors.

Source: GESAMP (1990).

Introduction

century it had grown to 2.5 billion and it currently stands at around 6 billion. Although higher figures have been widely quoted, in 1994, 37% of the world's population have been estimated to live within 100 km (62 mi) of the coast; coastal populations appear to be increasing at a higher rate than the population in general. Twelve of the twenty largest urban areas in the world are located within 160 km (100 mi) of the coast.

Sixthly, there has been the accelerated growth in science-based technology that has made it possible to discover new potentials as well as to exploit them, thus increasing the uses that must be accommodated by the oceans and the risks to which they are subjected. Many of the demands placed on the oceans are the direct consequence of technologies that only became available in the past few decades. While the application of new science-based technology to the oceans has greatly added to our understanding of the oceans as well as helped meet the basic needs of a rapidly growing population, it has become apparent that some of the uses have been ill-conceived and have contributed to the deterioration of the health of the oceans and unsustainable exploitation of marine resources.

Seventhly, the capacity of sovereign states to effectively govern the oceans has also been affected by the growth in criminal activity on the oceans, including the drugs trade, arms smuggling, illegal fishing, unauthorized waste disposal and piracy. When governments are unable to control activity at sea, they find it increasingly difficult to uphold public order on land.

The shift in our perception of the oceans brought about by these changes has also been reinforced by other developments. Dominant among these has been the recognition that the benefits derived from the use of the oceans and from the exploitation of marine resources accrue mainly to nations with the required scientific, technological and financial capacity. This has raised the issue of equity and highlighted the need for mechanisms that ensure that all nations are able to share in the benefits arising from the use and exploitation of the oceans. Many regard the oceans as providing a unique opportunity to develop innovative forms of governance that respond to the rights and needs of humankind, in particular those of poor and disadvantaged people.

Recognition of the importance of the oceans to global climate has also contributed to our changing perception. Although we refer to the oceans, suggesting the existence of self-contained entities, they together form a single global system that is in constant movement. This mass of water stores solar heat, and this capacity for storage is the great stabilizer of climate. This is changing due to global warming, the effects of which include sea-level rise, likely increases in the

Introduction

frequency and intensity of storms, changes in the location and abundance of fishing grounds, and changes in coastal ecosystems. As the main driving force for global climate, the oceans are of critical importance to the future well-being of the world's population.

OBSTACLES TO CHANGE

Efforts to establish forms of governance that make it possible to meet the many challenges posed by the oceans are confronted with numerous obstacles to change.

First, and most obviously, there are political obstacles to change, especially the resistance that can be expected from those who benefit disproportionately from current arrangements or from the absence of them. States with large navies are, for example, unlikely to be enthusiastic supporters of measures aimed at regulating the uses of the high seas when such regulations would impose restrictions on the activities of naval vessels. Similarly, nations possessing the technological capacity to exploit the newly discovered resources of the seabed can be expected to resist efforts to introduce mechanisms designed to enable states without such capacity to share in the benefits of exploitation.

Secondly, there are ideological barriers that shape perceptions of the need for change. In the economic sphere the dominant ideology is one that places a heavy reliance on the market and on the role of private enterprise and investment within a framework of deregulation and reduced government intervention. This ideology is not receptive to the establishment of the regulatory mechanisms that may be required for the oceans. It can be shown that markets provide a very inadequate response to many of the problems existing in the oceans, and that the absence of regulatory mechanisms has contributed to the abuse of the oceans and to the unsustainable exploitation of marine resources. Nor are markets, when left to their own devices, able to dispense answers to questions relating to peace, security, equity and justice.

Thirdly, there are barriers in the form of the behaviour of individuals. Although it is legitimate to refer to a growing awareness of the problems in the oceans, reflected in the increase of action groups and NGOs, this awareness does not yet permeate individual behaviour. Consumption patterns will remain incompatible with the sustainable use of marine resources until such time as individuals fully comprehend the consequences of the choices they make. At the other end of the spectrum, there are the hundreds of millions of poor people who are compelled for their survival to commit ecological destruction. Often the last link in a long chain of exploitative relations, they are left with little to exploit but nature in their daily struggle for survival, a situation that also links poverty with the state of health of the oceans.

Fourthly, the search for more effective forms of ocean governance has been rendered more complex by the increased number of nation states and by the need to ensure that systems of governance balance the needs and priorities of an increasingly heterogeneous community of nations. Only 86 States participated in the First Law of the Sea Conference held in 1958. By the time of the Third Conference, launched in 1973, the number of states had increased to 160, while today 185 nations negotiate on matters pertaining to the oceans. Not only has the number of negotiating parties more than doubled, the traditional negotiating 'blocks' that have facilitated negotiations on many other issues – such as disarmament, development, trade and investment – are much less in evidence in the case of the oceans. This inevitably has made the formation of consensus more difficult.

Fifthly, there are obstacles imposed by the complexity of the issues and by the realization that, in many important respects, our knowledge to effectively take them into account is still inadequate. It is difficult to forge a consensus on a way ahead when specialist opinion is divided, as it is in the case of some ocean problems and the importance of some marine potentials, or when the information required for informed decision-making does not exist.

Together, these obstacles hamper the development of the mechanisms required to regulate more effectively the uses of the oceans and to ensure that the benefits of the oceans are shared more equitably. In the meantime, the many demands on the oceans increase, pushing us closer to a situation that has been described as the 'tragedy of the commons'. This tragedy results from the exploitation of a common resource by individual users based on narrow economic motives until a point is reached when exploitation becomes unsustainable, ultimately dooming the commons to destruction. Experience tells us that, in such a situation, the only protection from irreversible damage is the development of arrangements that relate use rights to overall carrying capacity in an equitable manner.

SIGNS OF POSITIVE CHANGE

The above suggests that the task of developing appropriate systems of governance for the oceans will prove to be difficult. Many significant attempts have been made and are under way to adapt ocean governance to a situation of scarcity and conflict. The most comprehensive of these efforts has undoubtedly been the Third United Nations Conference on the Law of the Sea. Concluded in

Introduction

1982, after a process of negotiation that extended over almost a decade, the resulting Convention finally entered into force in 1994. Arguably one of the major achievements of intergovernmental negotiations this century, the Convention is one of the most elaborate treaties ever concluded and its negotiation succeeded, often as a result of the willingness of parties to travel uncharted paths, in reconciling the interests and positions of different groups of nations. There is no doubt that the Convention has helped to prevent anarchy in the oceans. Without it, the intensification of the uses of the sea, and the competing nature of many of them, would almost certainly have resulted in major conflicts about access to, and ownership of, resources. Although there are a number of areas in which the Convention fails to go far enough – such as security, resource management, and the preservation of biological diversity – it remains a remarkable agreement that establishes a foundation on which to build the future system of ocean governance. It is, accordingly, of the utmost importance that all States become parties to the Convention and that its provisions be fully and effectively implemented.

Efforts to shape systems of governance are also facilitated by a growing recognition of the importance of development that is not only economically and socially, but also environmentally, sustainable. An important turning point for acceptance of this position was the 1992 UN Conference on Environment and Development (UNCED) held in Rio de Janeiro. The Rio Declaration and Agenda 21 endorse the need for collective and concerted action to protect the global commons, that are defined to include the oceans, and Agenda 21 makes special reference to the importance of coastal zones. They also recommend the involvement in the decision-making process of all parties concerned and thus a more democratic process.

It is against this background of challenge and opportunity that the Independent World Commission on the Oceans has identified a number of directions for future action and debate, which will be elaborated in this report.

In the late twentieth century, peace is increasingly understood not just in military terms, and not just as the absence of conflict, but as a phenomenon encompassing economic development, social justice, environmental protection, democratization, disarmament and respect for human rights. These pillars of peace are inter-related and mutually reinforcing. Building peace and combating threats to peace in an interdependent world requires the full participation of every citizen, every nation, every continent. Governments, non-governmental organizations, private sector business and industries, academic institutions, trade unions and other members of civil society, are all players on the international stage. At the United Nations, all can come together in common cause to address today's complex global problems and to work towards shared goals.

Kofi Annan, UN Secretary-General
International Day of Peace, 16 September 1997

PROMOTING PEACE AND SECURITY IN THE OCEANS

1

This report will first consider problems related to 'the peaceful uses' of the oceans by governments, and their security forces, especially when operating outside of territorial waters. It also addresses a range of security concerns, including protection against threats directed at legitimate commercial and recreational uses. These threats may involve not only many types of illegal trade but also disregard for environmental standards, various types of unsafe practices and criminal violence. In this review of issues, the principal focus is on the minimization of recourse to force and on provisions that promote security for all legitimate users.

THE ORIGINS OF THE DEBATE

Concern for peace and security in the oceans is as old as the use of the oceans, with societies devising their own customs and practices aimed at providing required safeguards. However, the modern approach to peace and security in the oceans dates back to the first decade of the seventeenth century, when the Dutch East India Company recruited a young Dutch jurist, Hugo Grotius, to defend its seizure of a Portuguese treasure galleon when it passed through the Strait of Malacca.

The Portuguese contended that the sea, like the land, was subject to the exclusive dominion of sovereign states, a position that, at the time, was shared by such powers as Spain, Denmark, the Ottoman Empire, and the city states of Genoa and Venice.

'The sea, since it is as incapable of being seized as the air, cannot be attached to the possession of any particular nation.'

'every nation is free to travel to every other nation, and to trade with it.'

Hugo Grotius

The Dutch argued that the sea belonged to no one and, as such, could not be appropriated or made the subject of territorial claims. Grotius, in his short treatise *Mare Liberum* (1609), defended the Dutch position as follows: 'The sea, since it is as incapable of being seized as the air, cannot be attached to the possession of any particular nation' and, further, that an 'unimpeachable axiom of the Law of Nations... the spirit of which is self-evident and immutable, to wit: every nation is free to travel to every other nation, and to trade with it.' In articulating this position, Grotius was able to draw upon earlier thinking of Spanish theologians as well as on the liberal traditions of freedom of trade and navigation that existed at the time in other parts of the world, especially in Asia.

The Dutch won the legal argument and the narrow and short-sighted interpretation of 'freedom' was rejected. Had it prevailed, it would have undoubtedly encouraged nations to attempt to extend their territorial dominion over vast stretches of oceans and reinforced the reliance on naval power. This would have created a breeding ground for conflict and war and would have greatly curtailed opportunities to use the oceans for the common benefit of humanity.

The advocacy of the 'freedom of the sea' caused considerable controversy among jurists and diplomats at the time. However, as states increasingly came to recognize its many advantages, support for the freedom of the sea increased until it was generally accepted by the end of the seventeenth century. This acceptance has provided the keystone for public order on the oceans up to the present time.

The regulation of military activities on the oceans continues to be minimal, resting on such core ideas as *res communis* and freedom of navigation. However, two very general guidelines follow from the principle that the oceans are a shared resource: (i) all who use the sea are entitled to do so in peace and security; and (ii) all users share responsibility for the maintenance of peace and security.

'PEACEFUL USES':
CONCEPT, ISSUES AND PLAYERS

The concept of 'peaceful uses'

One fundamental concept implicitly underpins the notion of the peaceful uses of the oceans, namely that all peoples should benefit from their use. The concept also recognizes peace as being more than the absence of war, extending the notion of peace to include the idea of an equitable public order that governs all human activity. This broader notion of peace can be expressed differently by the insistence that the opposite of peace is not war but injustice.

The translation of the moral and ethical goal of peaceful uses into practical reality is the greatest single challenge to national governments, international organizations and civil society. It is also of increasing concern to those who use the oceans and have a stake in their future as well as those mandated to administer them.

Three central problems are immediately apparent. The first is one of definition. The concept of 'peaceful uses' means very different things to different people, and at present there is no consistent third-party mechanism or procedure available to resolve differences in interpretation. Leading naval powers interpret the concept to mean an endorsement of their traditional roles, whereas states that may feel threatened and are without strong navies understand it to mean the limitation of naval operations to strictly defensive missions.

The second problem relates to potential conflicts. Under the Law of the Sea Convention, governments have a duty to limit their maritime activities to peaceful uses. But governments of sovereign states also have a duty to provide security for their peoples as well as the global community. Possible conflicts between these two undertakings are far from imaginary.

The third problem relates to implementation and enforcement. Although a large number of organizations are actively involved in ocean affairs, there are at present no institutions for ocean governance at either the global or regional level that are able to deal effectively with peace and security issues. As a result, it is difficult to implement agreed norms and standards, to resolve disputes as they arise, and to provide mechanisms for the joint and sustainable development of marine resources.

Freedom of the seas

A primary motive for Grotius' defence of the concept of the freedom of the high seas was the desire to remove and prevent restrictions on international trade. The permissive regime fostered by this concept has certainly contributed greatly to the growth of international trade as well as of international investment and shipping. More recently, it has provided an impetus to international cooperation in such areas as tourism and recreation, scientific research, maritime safety, resource conservation, environmental protection, and law enforcement. Despite these positive developments, it must be recognized that the freedom of the seas, as both doctrine and practice, has many critics, especially in developing countries. Critics argue that the doctrine effectively paved the way for power politics, gunboat diplomacy and the colonial order, and that it has been used to give legitimacy to the ambitions and priorities of the strong while adding to the vulnerabilities of the weak. The challenge then is to find a balance between the positive and negative implications of the freedom of the seas.

Maintaining and extending the beneficial uses of the oceans is a goal that enjoys widespread support and is accorded a high priority. Proposals for restrictions that would hamper the pursuit of this goal should thus be viewed with caution. In charting a way ahead we should not lose sight of the importance of evolving a sense of community in relation to the oceans that is guided by the principle of equity and which works for the benefit of all. Nor should we lose sight of the fact that the freedom of the seas has often been abused. In this situation, the principle of the freedom of the seas must be understood to imply a recognition of the legitimate need for effective security, a respect for international law, and a new balance between the discretion of the strong and the vulnerability of the weak.

The concept of the freedom of the seas is in the process of being revisited, with the new approach taking account of the present requirements of peace, security at sea, and the need for the more sustainable use of the oceans and their resources. The substantive focus of this approach is provided by the responsible exercise of freedom and sovereign rights. However, it has been argued that this should not be understood exclusively in terms of the responsibilities of the major military and economic powers. Instead, the aim should

The challenge then is to find a balance between the positive and negative implications of the freedom of the seas.

be to pave the way for more advanced forms of cooperation in equitable ocean governance that involve all states. Only this will bring 'peaceful uses' as a guideline into conformity with both 'peace' as an ideal and 'international law' as a body of rights and duties.

The role of navies

All states are formally committed to the idea of using peaceful – essentially non-military – methods for settling ocean disputes and resolving conflicts, and they are also required to make comprehensive efforts to avoid war and military confrontation. Yet, contemporary international law, including the Law of the Sea Convention, allows maritime states wide discretion, under the 'peaceful uses' banner, to engage in traditional naval practices, including manoeuvres, forward deployments, threats and even forcible acts. These rights, and their exercise, have not been subjected to serious scrutiny for many decades; naval powers have consistently opposed this.

At the same time, much of what is presently known about the sea and many of the most significant ocean engineering advances have come about as a consequence of naval research and development for strategic, military and security applications. Most of these technological developments, along with masses of sensitive data, have traditionally been classified and have thus not been available for use in civilian scientific research, by private enterprise or by technologically less advanced nations. However, this situation is slowly changing. The end of the Cold War has brought about an encouraging willingness to release information and to make military technology available for civilian purposes. A formerly secret global underwater listening system (SOSUS) – developed for the US Navy to detect the presence and movements of submarines and other ocean traffic – is, for example, now being used by scientists to gain new insights into the movements and activities of sound-producing marine species, especially whales, as well as to obtain valuable information on seismic activity.

Non-governmental actors

Peace and security on the oceans does not only involve governments, their navies and other security forces. It also involves a wide range of non-governmental players, of which the scientific community, private enterprise and NGOs figure among the most important.

NGOs are playing an increasingly important and vociferous role in issues of peace and security. Many share a commitment to peace on the oceans and support demilitarization and denuclearization. Some have been primarily identified with campaigns to advocate the sustainable use of resources as well as the safeguarding of biodiversity. These NGOs typically regard 'peace' on the oceans as a matter of substantive duty and constraint – a position that contrasts with the one held by naval powers that appear to regard 'peaceful uses' as implying the absence of infringements on their freedom of action.

Although it is nation-states that negotiate and conclude treaties, NGOs have been able to influence public opinion and government positions in tangible ways. Achievements in these areas have been numerous and include the imposition of prohibitions on the testing of nuclear weapons on or under the ocean and the decision to keep Antarctica demilitarized and free from mineral exploitation.

Many activities in and on the oceans involve the assertion of claims by private enterprise. Some of these activities are based on narrow economic considerations and complicate the tasks of ensuring peaceful uses. They also raise new problems associated with the protection of legitimate activity taking place on the high seas.

Commercial corporations and consortia are increasingly able to deploy advanced technologies in the exploitation of the oceans and their mineral resources. Such technologies are usually protected by property rights and are capital-intensive, which restricts access for weaker and poorer states. Some of the technologies have negative impacts on traditional users of the oceans, who may be heavily dependent on relatively simple technologies for the localized exploitation of marine resources. Large modern corporate ventures in the oceans often resemble mega-science endeavours, and involve large investments that may include the construction of costly installations in or on the sea. These installations need to be protected by security forces against threats of vandalism, terrorism, piracy and possible predatory practices of competitors.

There is encouraging evidence that large corporations active in shipping and the exploitation of marine resources are willing to adopt longer-term and less narrow approaches to their activities. In some instances, this has included support for the preparation and implementation of voluntary codes of self-restraint in environmentally harmful behaviour. Such initiatives suggest a growing readiness on the part of private enterprise to make a commitment to environmental security. They also suggest that, rather than being an obstacle – the role in which it is traditionally cast – private enterprise is able to become a positive agent for change.

New threats to the oceans

All law-abiding states have an interest in suppressing illegal activities on the high seas. Historically, one of the first formal acts of international cooperation was a treaty in the nineteenth century to suppress the use of the oceans for the international carriage of slaves. This established the idea that navigational freedom should not be allowed to facilitate illegal trade and criminal practices. Given the growth of new forms of illegal trade, organized crime and piracy in some regions, the need to establish a balance between freedom and security is again becoming a challenge.

New forms of illegal trade include the transport of drugs, arms, clandestine immigrants, protected animal and plant species, toxic materials and nuclear wastes, as well as the dumping of environmentally harmful and hazardous substances banned under international agreements. The oceans are being used, for example, for an illegal trade in ozone-depleting chlorofluorocarbons (CFCs) which, in some countries, is today more profitable than trade in heroin or cocaine. Peace and security on the high seas are also threatened by international terrorism.

While it is in the obvious interest of all to ensure that the oceans do not become a zone of rampant criminality, the general absence of a regulatory presence makes it difficult to establish the required safeguards.

Historically, one of the first formal acts of international cooperation was a treaty in the nineteenth century to suppress the use of the oceans for the international carriage of slaves.

The growth of piracy and maritime violence

Although the seventeenth and eighteenth centuries were piracy's golden age, modern-day piracy is a serious and growing problem in several parts of the world. Today's pirates tend to operate out of small, fast boats, are frequently heavily armed, and often target large vessels carrying valuable cargo.

Location	Number of incidents					
	1992	1993	1994	1995	1996	1997
South-East Asia	63	16	38	71	124	91
Far East	7	69	32	47	17	19
Indian Sub-Continent	5	3	3	24	26	34
Americas	--	5	11	21	31	34
Africa	--	8	6	21	26	41
Rest of World	--	--	--	4	4	10
Not identified	31	2	--	--	--	--
Total	106	103	90	188	228	229

Moreover, fatalities associated with piracy increased during 1997 compared with the same period in the previous year, rising from 26 to 51 reported deaths. Finally, it is believed that the number of actual incidents of piracy is considerably higher than reported.

The International Maritime Bureau (IMB) defines piracy as an act of boarding any vessel with the intent to commit theft or any other crime with the intent or capability to use force in the furtherance of that act. (This definition of piracy differs from the one given in article 101 of the Law of the Sea Convention.)

Source: ICC International Maritime Bureau (1997).

Law enforcement and peacekeeping

Because the international community lacks effective mechanisms for law enforcement, the burden of providing security on the oceans has traditionally fallen on national navies and coastguards, although contributions have also been made by port security forces and even river police. On the high seas especially, ocean governance has to a considerable extent been taken over by leading maritime powers that purport to act as security agents for the world. However, the possible role of navies in ocean governance is controversial. While some regard it as a positive development, others are fearful that this regulatory arrangement is tilted heavily towards the protection of the interests of the rich and powerful.

The maintenance of the freedom of the seas must be balanced against the claims of coastal states. Under the Law of the Sea Convention, the authority of coastal states has been very much extended. Most have now established Exclusive Economic Zones (EEZs) extending up to 200 nautical miles from their 'baselines', and they have sovereign rights over the resources of their Continental Shelves that, in some cases, can extend even further. While these extensions of coastal state jurisdiction – the greatest change in the legal regime for the oceans in several centuries – primarily benefit individual countries, they also help to compensate for the absence of regulatory authority and enforcement mechanisms at sea.

However, the effects of these changes in the structure of authority remain uneven and uncertain. The governments of coastal states differ greatly in terms of their priorities and the methods used to pursue them, with differences reflecting technological and administrative capabilities as well as attitudes towards the oceans. If prudently exercised, the extended rights of coastal states could contribute to the security of nations and peoples in a sustainable manner as well as help to promote the goal of peaceful uses. However, it needs to be realized that navies are not subject to restrictions, even within the EEZ of a foreign state.

Maritime security is no longer – if indeed it ever was – the exclusive concern of the five major maritime nations, the permanent members of the UN Security Council. Ocean security concerns all nations, and an increasing number of them are making determined efforts to strengthen their capacity to uphold their legitimate interests at sea.

Naval activity and operations

During the last half-century, the conflict between the freedom of the high seas and national and international security has not really been taken up in international negotiations on the oceans. Consequently, the international legal community has not been invited to discuss systematically the subject of naval activity and the possible imposition of legal limits on the operations of naval powers. Nowhere was this more in evidence than in the negotiations during the Third UN Conference on the Law of the Sea. Although they covered virtually every aspect of ocean use, naval activities and operations were expressly excluded.

How can this omission be explained? At the time, the leading maritime states argued that this conference did not provide the proper setting for the discussion of naval rights and duties. They contended that special arrangements – where necessary and desirable – could and should be negotiated in separate fora. Examples of such arrangements include: the 1963 Limited Nuclear Test Ban Treaty (now superseded by the 1995 Comprehensive Test Ban Treaty); the 1971 Treaty on the Prohibition of the Emplacement of Nuclear Weapons and Other Weapons of Mass Destruction on the Seabed and the Oceanfloor and in the Subsoil Thereof; the 1973 Protocol to the 1972 US–USSR Agreement on the Prevention of Incidents on and over the High Seas, and the corresponding 1998 US–Chinese Agreement; and the 1995 indefinite extension of the Non-Proliferation Treaty and its review in the year 2000.

For centuries, major maritime powers have understood their security interests as requiring as much free use of the sea as possible. Ever since the principle of *mare liberum* was first articulated, discretionary reliance on naval power has never been legally challenged. Even now, the logic of military necessity, in times of peace or war, shapes the global role of navies.

The Law of the Sea Convention imposes on all states an overall obligation to limit their activities on and in the oceans to peaceful uses. The tacit understanding is that naval activities on the oceans in times of peace are to be considered peaceful uses, or preparations for individual and collective self-defence, and hence legal.

The emphasis placed on unrestricted freedom by naval powers and their own understanding of their security interests appears inconsistent with the promotion of peace and security on the oceans and at odds with the extended regulatory power of coastal states. While prevailing power realities seem to suggest that it may not be possible to severely restrict the role of navies in the near future, the issues of the freedom of the high seas and national and international security should be reinstated on the international agenda, so as to develop new arrangements that would benefit and protect the global community. Consideration also needs to be given to the role of international institutions in promoting peace-keeping operations at sea, working directly with national naval forces or through other appropriate arrangements.

UNCLOS and the peaceful uses of the seas

The commitment to 'peaceful uses of the seas' is expressed in Article 301 of the 1982 UN Convention on the Law of the Sea:

'In exercising their rights and performing their duties under this Convention, States Parties shall refrain from the threat or use of force against the territorial integrity or political independence of any State, or in any manner inconsistent with the principles of international law embodied in the Charter of the United Nations.'

In addition to this Article, the Convention refers elsewhere to 'peaceful purposes' as they relate to the high seas, the Area and marine scientific research.

Source: UN Convention on the Law of the Sea (1982).

Improvements in ocean security

The end of the Cold War has fundamentally altered the context in which the quest for peace and security on the oceans is taking place. The biggest and most welcome change brought about by the removal of tensions is the diminished likelihood of global warfare. During the Cold War, priority was given to nuclear deterrence, maintained at sea by submarines which were supposed to deter a surprise attack by an adversary. Despite the ending of the Cold War, force structures and even deployment patterns have yet to be adapted to the new reality of greatly reduced strategic conflict among the naval superpowers.

Notwithstanding current inertia, there have been important improvements in ocean security. There is far less prospect today of hostile incidents at sea and, should they occur, of their uncontrolled escalation. Given prevailing circumstances, a major mission of global naval forces is to respond to regional threats, as most notably in the various phases of the 1990–1991 Gulf crisis. A remarkable feature of this recent pattern is the degree to which the UN Security Council has mandated important military undertakings when in line with new power realities. During the Cold War, superpower rivalries effectively prevented the UN from taking decisions on such matters, since the superpowers not only took divergent positions on the conflicts but also vetoed proposals for dealing with them.

ENHANCING PEACE AND SECURITY: THE WAY AHEAD

The international community has yet to develop a major institutional capacity for promoting and regulating peace and security on the oceans. The military aspects of peace and security, for example, depend almost exclusively on the prudence, self-restraint and capabilities of the leading naval powers, as well as on the acquiescence of other states. It is a highly voluntaristic type of world order based on a combination of self-help and the benevolence and effectiveness of a few governments. The question has to be raised whether these voluntaristic arrangements can be expected to work in the next century, and whether they are responsive, at the global level, to democratic values and considerations of equity.

The international community has yet to develop a major institutional capacity for promoting and regulating peace and security on the oceans.

Possible initiatives for a more affirmative approach to the promotion of peace and security on the oceans are presented below with a view to promoting discussion and action on a crucially important issue of ocean governance.

- **Treating the 'high seas' as a public trust**

Peace and security on the oceans would be greatly facilitated by the effective implementation, by all states, of the Law of the Sea Convention and its implementing agreements. It would be further enhanced by new initiatives that give tangible expression to the recognition that the high seas cannot be appropriated by any state and should be reserved for beneficial uses. Given this acceptance, the most appropriate concept for the high seas seems to be that of 'trusteeship' or 'stewardship'.

Trusteeship is a well-established common law concept. It has historical parallels in a number of legal systems, including the *fiducia* and *fideicommissum* in ancient Roman law, the *waqf* in Islamic law, and the *moramati* in African customary law. In essence, it may be defined as a fiduciary relationship with regard to specific resources (the 'trust'), imposing on those who hold the trust (the 'trustees') certain equitable obligations to manage the trust for the benefit of others (the 'beneficiaries'). The concept has been applied by philosophers of different ideological persuasions – for example, Thomas Aquinas (*Summa Theologica*), John Locke (*Social Contract*) and Karl Marx (*Das Kapital*) – who have postulated the duty to manage the Earth for the benefit of future generations. The 'public trust' doctrine, developed by American jurisprudence since 1892, similarly confers on governments a permanent duty to manage natural resources for the benefit of all. States as trustees are subject to certain fiduciary obligations ('public interest norms') in managing and using the natural resources concerned on behalf of all beneficiaries and in the interests of future generations according to recognized standards of care. Such standards can be derived from analogies with national trust law.

...the most appropriate concept for the high seas seems to be that of 'trusteeship' or 'stewardship'.

Basic fiduciary rules

(i) A trust is not a legal entity; it is not the bearer of rights or the subject of duties.
(ii) Rights to the resources subject to a trust are divided between the trustee(s) and the beneficiaries.
(iii) A trustee must manage the trust solely in the interest of the beneficiaries.
(iv) A trustee must earmark trust resources as such, so as to ensure the integrity of the trust and avoid their confusion with his own resources.
(v) In managing the trust, a trustee must use reasonable care and skill, and avoid unreasonable risks.
(vi) A trustee is accountable to the beneficiaries for his management of the trust.

Source: Adapted from Gold (1978).

In the application of the concept of trusteeship to the 'high seas', much remains to be clarified. Procedural mechanisms would need to be developed as would arrangements for dealing with supervision and non-compliance with the terms of the trust. The most appropriate arrangement would be the reconstruction of the UN Trusteeship Council with a composition that reflects the international community as a whole and with a mandate that makes it possible for it to exercise trusteeship functions for the 'high seas'. However, such a reconstitution would probably require the revision of the UN Charter and should thus be regarded as a longer-term option. At present, the UN General Assembly is the only fully representative body for the purpose of global trusteeship.

● **Promoting peaceful resolution of conflicting maritime claims**

Numerous regional and sub-regional disputes exist over sovereignty claims. Many involve islands that possess either a symbolic or an economic value. While some have been the subject of disputes for more than a century and others are the legacy of the Cold War, the Law of the Sea Convention's provisions for extended coastal state jurisdiction have created a situation leading in some cases to conflicts. If left to fester, these conflicts are likely to further heighten tensions and

Major maritime cases brought before the International Court of Justice

From 1946 up to 1998, 17 major maritime cases were brought before the International Court of Justice at The Hague, showing a steady increase: While there were only 3 cases during the first 20 years (1946–1965), the number doubled during the next 20 years (1966–1985) and may double once more, with 4 cases already decided and 4 more pending during the 12 years since 1986.

- Corfu Channel (United Kingdom v. Albania), 1949.
- Fisheries (United Kingdom v. Norway), 1951.
- Minquiers and Ecrehos (France v. United Kingdom), 1953.
- North Sea Continental Shelf (Federal Republic of Germany v. Denmark; Federal Republic of Germany v. Netherlands), 1969.
- Fisheries Jurisdiction (United Kingdom v. Iceland; Federal Republic of Germany v. Iceland), 1973.
- Aegean Sea Continental Shelf (Greece v. Turkey), 1978.
- Continental Shelf (Tunisia v. Libyan Arab Jamahiriya), 1982 and 1985.
- Delimitation of the Maritime Boundary in the Gulf of Maine Area (Canada v. United States of America), 1984.
- Continental Shelf (Libyan Arab Jamahiriya v. Malta), 1985.
- Land, Island and Maritime Frontier Dispute (El Salvador v. Honduras; Nicaragua intervening), *Partial Judgement*, 1992.
- Maritime Delimitation in the Area Between Greenland and Jan Mayen (Denmark v. Norway), 1993.
- Maritime Delimitation and Territorial Questions (Qatar v. Bahrain), *Jurisdiction and Admissibility*, 1994.
- East Timor (Portugal v. Australia), 1995.

Pending cases (interim orders):
- Maritime Delimitation (Guinea-Bissau v. Senegal), 1991–
- Oil Platforms (Islamic Republic of Iran v. United States of America), 1992–
- Land and Maritime Boundary (Cameroon v. Nigeria), 1994–
- Fisheries Jurisdiction (Spain v. Canada), 1995–

Source: International Court of Justice (1998).

The
active
involvement
of navies
in law
enforcement...
could serve
to make
the oceans
safer
for the
global
community.

could even result in military confrontation. This could give rise to regional crises, with the possibility, in some instances, that extra-regional states, aligned to one or both sides, would become involved. Such spirals of conflict are unpredictable and would certainly damage efforts to promote peace and security in the oceans.

In such cases, it would be particularly desirable to freeze some or all sovereign claims, for an indefinite period or at least for a number of years. The freezing of claims would be followed by an intergovernmental approach and recourse to procedures that have already averted several conflicts.

There are various overlapping ways to resolve regional disputes over the oceans: negotiations, diplomacy, international adjudication, arbitration, freezing of claims, or joint development of off-shore resources. In many settings, joint development seems politically feasible, practically useful and legally sound. From a legal point of view, a body of rules and practices on joint development in the setting of contested ocean claims is already evolving into a body of customary international law.

There are now 17 joint development schemes in operation around the world. Until such time as international law develops a clearer approach to the resolution of regional seas conflicts, especially in the settling of claims involving uninhabited islets and rocks, provisional agreements seem to offer the most promising avenue for the future.

● **Reorientating the security roles of navies and other maritime security forces**

While obviously a delicate issue, the active involvement of navies in law enforcement in the face of new threats – such as those emanating from illegal trade, clandestine transport of persons, eco-crime, congested shipping lanes, piracy, and terrorism – could serve to make the oceans safer for the global community. This suggestion should not be interpreted as a plea for larger navies, which would be inconsistent with the promotion of peace and security in the oceans and would certainly be unacceptable in some regions. Rather, it is a suggestion that calls for a reorientation of naval responsibilities in conformity with present international law and in ways that could contribute positively to the demilitarization of the oceans.

Action in this area would need to be translated into specific contexts. It could, for example, be based on a regional approach in which navies from neighbouring countries agree to pool their resources and to undertake joint monitoring and surveillance. It could also form part of a more formalized and broader-based programme of regional cooperation in respect of the joint management of coastal waters and off-shore resources.

Efforts to reorientate naval responsibilities should include the creation and training of specialized naval units, mandated and equipped to undertake policing tasks supported, whenever possible, by the use of ocean observation satellites. Where navies are reluctant to depart from tradition, the capability of national coast guards and other security forces could be appropriately strengthened.

Even a partial reorientation along the lines proposed would have significant implications for the future funding, planning and procurement programmes of navies. It would also allow for a new allocation of functions between navies and other maritime security forces.

● Moving by stages towards nuclear disarmament in the oceans

There is a strong public aversion to nuclear force and the presence of nuclear weaponry on the high seas. This presence was explained during the Cold War by the doctrine of mutual deterrence. However, given changed political realities, there is today no form of deterrence that cannot be carried out by non-nuclear naval and other military forces.

The risks associated with nuclear accidents – of which several have already occurred – as well as the dangers created by the disposal of radioactive wastes at sea provide further justification for the pursuit of denuclearization. The goal of denuclearization was strongly affirmed by the 1996 Advisory Opinion of the International Court of Justice in response to a question put by the UN General Assembly regarding 'The Legality of the Threat or Use of Nuclear Weapons'. It was further endorsed by the nuclear disarmament call of the Canberra Commission on the Abolition of Nuclear Weapons, the members of which included prominent national security officials drawn from nuclear weapons states. The prevention of further proliferation of nuclear weaponry would be facilitated

by new initiatives in the general direction of nuclear disarmament, including the eventual removal of all weaponry of mass destruction from the high seas.

Efforts to achieve a gradual denuclearization of the oceans should also include the following:

● Encouraging phased disarmament negotiations among the states with nuclear weapons with the aim of reducing and eventually eliminating nuclear weapons from the high seas, and the discharge, in good faith, of their legal and moral obligation to achieve complete nuclear disarmament.

● Extending the establishment of nuclear-free zones in regional seas through agreements among states and the encouragement of universal observance of the regimes established. New initiatives would build upon the Treaty of Tlatelolco (1967), which commits parties not to test, deploy or use nuclear weapons in Latin America, and on the Treaty of Raratonga (1985), which prohibits parties from testing, developing and deploying nuclear weapons in the Pacific.

These zones have so far not been respected by nuclear states because they would violate the freedom of navigation, the argument used in the refusal to sign the agreement establishing the South East Asia Nuclear-Free Zone. As matters now stand, nuclear-free zones tend to be little more than a declaration of the policies adopted by non-nuclear states in the region concerned. Nevertheless, they contribute to the formation of a regional consensus on denuclearization and on the avoidance of nuclear weaponry.

● **Ensuring observance of environmental and other legal standards by warships and military activities**

Sovereign immunity for warships and military activities would need to be qualified in the future, especially to permit the control of activities that pose serious threats to the environment, such as the release of radioactive materials and other ultra-hazardous substances. Given the importance of raising standards of law enforcement on the oceans, it is now indeed legitimate to question the extent of such immunity.

Although revolutionary changes have taken place in military activities on, under and over, the oceans, the law of naval warfare has not been revised for almost a century. This appears an opportune time for new law-making initiatives, the importance of which is underscored by the growth of overlapping and contradictory claims to the oceans and their resources, with their potential for incidents that pose a threat to peace.

The San Remo Manual of International Law Applicable to Armed Conflicts at Sea, drafted by a group of legal and naval experts in the period 1988–1994, provides a useful basis for updating the law of naval warfare. The need for legal reform has been recognized in a very special context in the form of the draft Protocol on Naval Mines, submitted by Sweden in 1994 to the Group of Government Experts established to prepare the review conference of the 1980 Convention on Certain Conventional Weapons.

The most appropriate step would be to encourage the UN, or some other appropriate international body, to organize a preparatory meeting, as soon as practicable, to consider which aspects of the global law of naval warfare are most in need of updating. The ultimate purpose of such a preparatory meeting would be to convene an international conference of governments and other important users of the oceans early in the next century. The conference might be held at The Hague so as to recall the historic peace conferences held on this and similar subjects in 1899 and 1907. To facilitate this undertaking, the International Law Commission could be requested to initiate the process by the preparation of background materials. The main naval powers could also be encouraged to convene their own conference for the purpose of updating the law of naval warfare and related activities. However, preference should be given to a process that extends the new law of naval warfare to the entire community of nations.

If the idea of such a conference fails to receive the support of the intergovernmental system, consideration should be given to the possibility of inviting NGOs to include the issue of naval warfare in their 1999 'Hague Appeal for Peace' and to formulate recommendations, that could include draft treaty proposals. Regional initiatives along similar lines might also be considered to complement naval law-making at a global level.

> Although revolutionary changes have taken place in military activities on, under and over, the oceans,the law of naval warfare has not been revised for almost a century.

● Preparation of a Report on Peace and Security in the Oceans in the 21st Century

Peace and security in the oceans raises many issues that are not only of crucial importance for the future of the oceans but also politically sensitive. They need to be examined systematically, both individually and in relation to each other, and in far more detail than is possible in this report. Consideration should accordingly be given to the preparation of a Report on Peace and Security in the Oceans in the 21st Century that looks in detail at the different issues and the relationships between them.

The proposed report could build upon the 1992 *An Agenda for Peace* submitted by the UN Secretary-General to the Security Council, extending the thinking to the oceans. The report could also be used to initiate action related to several other proposals contained in this report, such as steps that could result in treating the high seas as a public trust and preparations for the proposed review of the law of naval warfare.

In the preparation of such a report, it must be borne in mind that all states, including land-locked states, have security interests that are affected by the oceans. These interests relate to the safety of navigation and commercial relations, recreational and scientific uses of the oceans, the assurance that marine foodstuffs are fit for human consumption, and the suppression of transnational crime and terrorism. The requirements for maintaining peace and security in the oceans also need to be interpreted in the light of global and regional interests in promoting sustainable development, and of the special needs of developing countries. The report should also take into account the likelihood of serious territorial disputes involving islands and access to resources.

The proposals made above for increasing security at sea will help foster peace on land. It is today widely accepted that real peace is much more than the absence of conflict and war. Positive peace is a peace imbued with equity and justice and which establishes the conditions required to enhance sustainable social and economic progress. It is a peace that creates new opportunities and perspectives for those who, through an accident of birth, are condemned to a life of unremitting struggle and to a premature death. The debate on equity has so far largely ignored the contribution that can be made by the oceans. It is to this contribution that we will now turn.

Consideration should be given to the preparation of a Report on Peace and Security in the Oceans in the 21st Century ...

A true system of politics cannot take a single step without first paying tribute to morality.

Immanuel Kant

All states and all people shall cooperate in the essential task of eradicating poverty as an indispensable requirement for sustainable development, in order to decrease disparities in standards of living and better meet the needs of the majority of the people of the world.

Rio Declaration on Environment and Development, Principle 5, 1992

THE QUEST FOR
EQUITY IN THE OCEANS

2

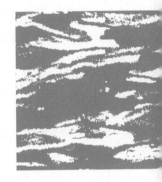

The consideration of equity in this report stands on three premises. First, systems of governance and management for the oceans must display a special sensitivity to the needs and requirements of groups and individuals who are disadvantaged by geography, by economic and social circumstances, and by their adherence to traditional methods of resource exploitation. Secondly, the oceans should be seen as a domain where institutions and arrangements should contribute to accelerating the pace of social and economic development in the developing countries, including those which are still without the capacity to use and benefit from the resources of the sea and those which do not have access to it by reasons of their geography. And, thirdly, systems of governance and management must recognize the legitimate interests of future generations, providing for arrangements through which the interests of the unborn can be articulated and safeguarded.

These premises point to the need for a fresh approach to the debate on equity in the oceans. This approach must give expression to the notion of human solidarity – a solidarity that engages all people while being protective of differences based on outlook, religion and culture. The approach must be guided by a sense of moral purpose and of longer-term goals. It must be rooted in arrangements that serve the common good and respond to the interests and needs of different users of the oceans, incorporating the overarching imperative of sustainable development. In effect, the achievement of more equitable outcomes depends on the fashioning of 'win/win' situations that reconcile different values and interests, and that rely on persuasion, mediation and compromise rather than on coercion, exclusive claims of right, and the dominion of technological advantage.

THE EQUITY ARGUMENT

This report builds upon the emphasis on equity embodied in Chapter 17 of Agenda 21. It acknowledges the close relationship existing between deprived segments of national populations and conditions in coastal areas, and it commits governments to manage coastal regions and off-shore resources with the conscious aim of alleviating poverty. As such, it establishes a foundation for making the public order in the oceans more equitable.

The oceans must be regarded as a common resource to be used and managed in the interests of all people. In this respect, 'all people' should not be considered as an abstraction but rather as a legitimate expression of the rich diversity inherent in the human adventure. This report affirms a belief in the enduring value of these differences, including those associated with traditional customs and lifestyles.

This is an approach that, at present, clashes with a dominant ideology that stresses the importance of markets and short-term gains over longer-term rewards. However, the commitment to 'fairness' and 'justice' cannot be entrusted to the unregulated behaviour of the market place. There is no doubt that the global market place has facilitated the rapid expansion of international trade and investment. From the perspective of equity, it is particularly notable that some developing countries have achieved impressive rates of growth that have translated into tangible improvements in standards of living. However, the impressive progress made by some countries only serves to highlight the stagnation in many others and, for many millions of people, the last two decades have witnessed declines in standards of living and an intensification of the daily struggle for survival.

According to UNDP's 1997 *Human Development Report*, 'the ratio of the income of the top 20% (of the world's population) to that of the poorest 20% rose from 30 to 1 in 1960, to 61 to 1 in 1991 – and to a startling new high of 78 to 1 in 1994.' From the equity perspective, this rise must be deemed unacceptable. Market forces, by themselves, will be unable to correct this situation.

The general efficacy of market forces is no excuse for ignoring their shortcomings or the damage they are able to inflict on human beings, the environment and the oceans. The harm is most in evidence when unsustainable practices are relied upon to achieve short-term profits, and when social groups or even whole societies are prevented from participating fully

> The oceans
> must be
> regarded as a
> common
> resource
> to be used
> and managed
> in the interests
> of all people.

in markets by a variety of constraints. These can include the lack of formal education and training, poorly developed capacities to exploit comparative advantages, or geographical disadvantage.

This situation forms a challenge that must be addressed by efforts to realize equity in the oceans. As formidable as it is, it is not the only challenge. The fundamental problem in the oceans is one of ensuring that various uses are reconciled with the imperatives of long-term sustainability and that gains derived from ocean resources are shared in a manner that benefits all nations and all people. The oceans need to be protected and developed on behalf of humankind, rather than sacrificed to expediency and short-term gain. Efforts are required to overcome some of the barriers that stem from insufficient scientific and technological capacities. Such ethical aspirations cannot be achieved immediately, but their articulation is important so that socially vulnerable groups can have points of reference by which to challenge policies and behaviour, whether of private or of public sector bodies, and produce a better balance between the operations of the market in the oceans and the well-being of people.

There are also new vistas of cooperation with the private sector. A number of transnational corporations – for instance, those associated in the World Business Council on Sustainable Development – are coming to accept that not only their public image but also their longer-term profitability will be enhanced if their corporate strategies embody a concern for sustainable

Marine Stewardship Council (MSC)

This independent, non-governmental, non-profit body is the result of a partnership between the World Wide Fund for Nature (WWF) and Unilever. Its purpose is to establish, on the basis of a broad set of standards for sustainable fishing, a system of certification for individual fisheries that are considered to follow good practice. Seafood companies will be encouraged to join sustainable 'buyers' groups' and purchase only fish products marked with a logo indicating that they come from certified fisheries. Independent, accredited certifying firms will certify only fisheries meeting the standards, thereby permitting consumers to select fish products that they know come from sustainable, well-managed sources.

Source: World Wide Fund for Nature (1997).

development. There is also growing evidence of a readiness to transform adversarial relations with environmental NGOs into partnership relations, with the cooperation between the World Wide Fund for Nature and Unilever on marine stewardship serving as a case in point.

It is encouraging that a number of major transnational corporations are pledging themselves to a stakeholder approach to sustainable development, with strategies that, for the first time, embody a real commitment to poverty alleviation in the developing countries in which they operate. Numerous multilateral and bilateral agencies, led by the World Bank, have also adopted policies that display greater sensitivity to the negative impacts of structural adjustment and processes of globalization.

EQUITY RECONSIDERED IN THE SETTING OF THE OCEANS

Until recently, concerns about the use and abuse of the oceans were not perceived as raising fundamental issues of equity. Equity was largely understood in terms of maintaining the freedom of the seas, which was assumed to be in the interest of, and beneficial to, all states and all people. A combination of factors – offshore hydrocarbon exploitation, overfishing, resource scarcities, resource conflicts and pollution among them – contributed to the realization, especially in developing countries, that the traditional understanding of equity was a very inadequate one and was in need of redefinition. The search for new approaches took two essential forms.

The first was the extension of coastal state jurisdiction, especially by Latin American countries during the 1950s and 1960s. They, and others, challenged the view that the 'freedom of the sea' was favourable to all, arguing instead that the freedom worked to the advantage of nations having the economic and technological capacities required to exploit the resources of the sea and to the disadvantage of those that did not. Arguments advanced by the developing countries in favour of extended coastal state jurisdiction were initially resisted by the industrialized countries. However, it became increasingly apparent from the arguments advanced that many of these countries stood to gain. An agreement was eventually reached that provided for the extension of coastal state jurisdiction over marine resources, mainly in the form of 200 nautical mile EEZs. The economic significance of the decision to establish EEZs is evident from the fact that they account for an estimated 8% of the earth's surface, 25% of global primary productivity and 90% of the world's fish catch. Moreover, as pointed out in Chapter 4, coastal marine environments encompassed by these EEZs, according to a recent estimate, form 43% of the value of the world's 'ecosystem services' which further underscores the global need for effective environmental protection in the EEZs.

EEZs have enormous implications for the quest for equity in the oceans. They mean that access to resources is no longer solely determined by capacities to exploit them, as they were under the 'freedom of the seas'. Under the provisions of the Law of the Sea Convention, coastal states enjoy exclusive rights to the exploitation of all living and non-living resources in their EEZs. This is a fundamental change that gives real expression to both equity and the legitimate aspirations of the developing countries.

Common heritage of mankind

Article 136

The Area and its resources are the common heritage of mankind.
(Article 1.1(1): "Area" means the seabed and ocean floor and subsoil thereof, beyond the limits of national jurisdiction.)

Article 137: Legal status of the Area and its resources

1. No State shall claim or exercise sovereignty or sovereign rights over any part of the Area or its resources, nor shall any State or natural or juridical person appropriate any part thereof. No such claim or exercise of sovereignty or sovereign rights nor such appropriation shall be recognized.

2. All rights in the resources of the Area are vested in mankind as a whole, on whose behalf the Authority shall act. These resources are not subject to alienation. The minerals recovered from the Area, however, may only be alienated in accordance with this Part and the rules, regulations and procedures of the Authority.

3. No State or natural or juridical person shall claim, acquire or exercise rights with respect to the minerals recovered from the Area except in accordance with this Part. Otherwise, no such claim, acquisition or exercise of such rights shall be recognized.

Source: UN Convention on the Law of the Sea (1982).

59

The second major development was the emergence of a revolutionary new concept, initially articulated by Arvid Pardo in 1967, asserting that the resources of the seabed beyond national jurisdiction should be regarded as the 'common heritage of mankind'. As a common heritage, these resources could not be appropriated by any nation but should be used for peaceful purposes and managed in the interests of all, including future generations. Originally conceived as having a potentially wider application, the concept was the subject of intense debate and there was considerable opposition, mainly from within nations with the technological capacity required to exploit the resources of the seabed. Eventually, the concept was applied to the seabed beyond national jurisdiction. The translation of this concept, as initially envisaged, into actual practice was reduced in scope with the adoption of the 1994 implementing agreement. However, since it still requires that all nations should benefit from the use of the 'common heritage of mankind', even in its reduced application, the concept has major implications for the pursuit of equity in the oceans.

These two broad normative developments are of great importance. Their significance is thrown into sharper relief by emerging evidence of the readiness of some powerful nations to question processes that are shaping the present international economic system, with their strong reliance on market forces.

An apparent indication of the change in thinking taking place was provided by the address of the US President to the 52nd UN General Assembly (1997): 'The United Nations could play a crucial role in making sure that as the global economy creates greater wealth, it does not produce growing disparities between the haves and have-nots, or threaten the global environment – our common home.' Similar sentiments have been expressed by other Western leaders and, when viewed together, they could be interpreted as implying that there is still a place for morality in politics. They could also be interpreted as an invitation to examine the extent to which globalizing pressures are accentuating income disparities, both between and within countries, and are contributing to the destruction of the environment. By implication, they may also suggest a willingness to consider more favourably proposals that may be formulated for addressing the situation.

There seems to be some movement towards a redress of the ideological balance, although not through the revival of old methods. Appeals to the governments of rich countries to provide direct assistance to those most affected by the negative consequences of globalization are unlikely to meet with much success, and little reliance can be placed on institutional innovations designed to alleviate the worst manifestations of poverty and inequality.

EQUITY CHALLENGES IN THE OCEANS

Progress towards greater equity in the oceans will be facilitated by the identification of groups that are particularly vulnerable to negative trends or are particularly disadvantaged by current patterns of use and abuse and are thus in need of protective regimes. This is an identification that can extend to communities, peoples, and nations.

A case could be made for the argument that vulnerabilities extend to the whole of humanity, since everyone is affected negatively, directly or indirectly, by such problems as 'dying seas', global warming, growing resource scarcities, and increasing levels of air and water pollution. However, some are in a better position to protect themselves from adversity than others, which is why we choose to draw attention to the predicament of the following groups.

Indigenous peoples. Indigenous peoples with strong links to the sea figure prominently on the list of vulnerable groups. For such peoples, access to and use of the sea is not only essential for their livelihoods but also gives meaning to their lives – a meaning which may have evolved over many centuries and in which behaviour and value systems cannot be explained without reference to relationships with the sea. The pressures on the very right to exist of indigenous peoples in many parts of the world are increasing as demands on the coastal zone and in-shore waters grow inexorably. Examples of special measures and arrangements designed to protect them from these pressures, to retain their distinct identities, and to safeguard their rights in maintaining access to the sea are conspicuously absent. The 1994 Draft Declaration on the Rights of Indigenous Peoples, concluded under the auspices of the UN, confirms the seriousness of the problem and outlines the elements of more a positive approach. Such an approach is in line with the 1992 Rio Declaration, which identifies indigenous peoples as one of the groups requiring special attention.

Traditional fishing communities. The challenge here is, in many respects, similar to the one posed by the vulnerabilities of indigenous peoples, although they may be less severe since they seldom include direct

threats of cultural extinction. They are more manifest in the depletion or even destruction of the fishing grounds on which the livelihood of traditional fishing communities depends, by insensitive and ill-considered on-shore development, pollution of the coastal zone, and by modern fishing vessels. Those vessels, often of distant-water fishing fleets, make use of sophisticated, capital-intensive technologies. Coastal authorities may be helpless to control them or may actually encourage them. Conflicts may also arise when traditional fishing communities harvest species, which – as a consequence of over-fishing by modern vessels – have been made the subject of fishing prohibitions.

There is more at stake than the protection of these peoples. The knowledge, experience and outlook of many traditional and indigenous peoples embody a deep wisdom about the human condition, and their survival skills may have a relevance to people everywhere, especially at a time in which the world is running up against a variety of ecological limits.

Coastal populations. A broader group that should be identified for priority action is the population of coastal zones. Many of the world's poor live close to the coast and are affected, directly and indirectly, by the deterioration that is taking place in coastal zones. In some parts of the world, the poor are compelled, in the absence of other alternatives, to occupy areas that make them particularly vulnerable to natural disasters and other hazards. Improvements in conditions of coastal zones could therefore be expected to bring immediate as well as longer-term benefits to some of the most disadvantaged sectors of society.

Small island countries. Vulnerabilities do not only extend to individuals and communities, they can also extend to whole nations. Nowhere is this more so than in the case of small island countries. While many of these countries figure prominently as major beneficiaries of the Law of the Sea Convention, with its provisions for greatly extended coastal state jurisdiction, few have the capacity at present to exploit and manage the resources to which they have exclusive rights. Because these countries are usually far more dependent upon the marine sector for their growth and development than other countries, the deterioration in the quality of the seas not only makes it more difficult to meet basic needs but also undermines the very economic viability of the nation. Many small island countries are low coral atolls, making them particularly vulnerable to sea-level rise. An increase in sea-level of about 50 cm (19.7 in) from the present to 2100, as projected in the 1995 Second Assessment of the Intergovernmental Panel on Climate Change (IPCC), will be sufficient to render many of them unsafe or unfit for human habitation, suggesting that it is not only vulnerable groups but also whole nations that are threatened with extinction.

Land-locked and geographically disadvantaged nations. The position that the oceans are a common domain and a common resource that should be used and managed for the benefit of all, means that it is not only coastal states and their populations that should figure as the beneficiaries of the oceans. Of the 185 member States of the UN, there are around 40 countries – most of them developing countries, all together they have a population of around 350 million people – that have no border with the sea. At present, they derive little direct and indirect benefit from the exploitation of the oceans. The hope of such countries that they would benefit from the provisions established by the Law of the Sea Convention has so far proved to be unfulfilled. Although the Convention affirms that land-locked and geographically disadvantaged countries should receive special attention, this has not happened. More than 15 years after the adoption of the Convention, measures aimed at enabling this group of countries to benefit from the uses of the sea and its resources are still not in place.

NORMATIVE FOUNDATIONS FOR EQUITY IN THE OCEANS

There exists ample legal and moral authority to establish the normative foundations on which to build greater equity in the oceans and to translate the principle of solidarity into practical policy. Most relevant perhaps is the language to be found in the Preamble to the Law of the Sea Convention, setting forth general directives that are expressed operationally in provisions distributed throughout the Convention. Among the goals affirmed is 'the equitable and efficient utilization' of ocean resources, with the Preamble observing that the goals pursued should 'contribute to the realization of a just and equitable international economic order which takes into account the interests and needs of mankind as a whole and, in particular, the special interests and needs of developing countries, whether coastal or land-locked'.

The Preamble goes on to affirm that the resources of the deep seabed and ocean floor are to be regarded as 'the common heritage of mankind' and should be used and managed 'for the benefit of mankind as a whole'. Elsewhere, the Preamble asserts that the oceans should contribute to 'the economic and social advancement of all peoples of the world' as well as to 'peace, security, co-operation and friendly relations among all nations in conformity with the principles of justice and equal rights'.

The 1992 Rio Declaration affirms similar principles. Several are of general applicability and should be regarded as indicative of the widespread support given by governments to equity in relation to development. Principle 3 of the Declaration asserts that 'The right of development must be fulfilled so as to equitably meet developmental and environmental needs of present and future generations'. Principle 5 affirms that 'All States and all people shall cooperate in the essential task of eradicating poverty as an indispensable require-ment for sustainable development, in order to decrease the disparities in standards of living and better meet the needs of the majority of the people of the world'.

Principle 6 insists that 'special priority' should be given to the needs of 'the least developed and those most environmentally vulnerable'. Also relevant is the emphasis placed on the importance of enlisting the participation of all people in support of these goals, with special consideration being given to the importance of women (Principle 20), youth (Principle 21), and indigenous people (Principle 22). It would be useful to adapt these principles to the specific conditions of coastal communities. A code of conduct for the use and administration of coastal waters would be particularly helpful, in the light of objective 17.5 of Agenda 21, for the 'integrated management and sustainable development of coastal areas and the marine environment', including EEZs.

The positions contained in these declarations and statements have been endorsed by the community of nations. They help to establish the consensus and normative foundations required to realize greater equity in the oceans.

Set out below are several areas in which positive action would contribute significantly to the realization of equity in the oceans. Many of the proposals take specifically into account the provisions and the understanding already reached between industrialized and developing countries, as embodied in the Convention. It has become apparent that some parts of the Convention are not being fully implemented. It must be recognized that the main value of the Convention to the community of nations resides in its comprehensive character. It can be fairly characterized as a 'bargain' that was struck between different groups of countries. For the technologically more advanced countries to fail to implement some parts of the Convention is to overlook the extent to which they sought and obtained the agreement of other groups of countries for other parts of the Convention.

This applies especially to the provisions contained in the Convention's Part XIV on the Development and Transfer of Marine Technology. Articles 266–278 set forth a broad programme of commitments, including the duty to promote actively the sharing of technology for the use and enjoyment of all marine resources. The promises contained in the Convention are explicitly extended to apply to training, research and technology transfer. A particularly innovative feature is the call in Articles 275–277 for the establishment of National and Regional Marine Scientific and Technological Centres.

It is against this background that many of the following recommendations are made.

Management of the coastal zone

The importance of the coastal zone requires more vigorous attempts on the part of national and local governments to regulate activity, reconcile conflicting uses, address sources of land-based degradation and pollution of the marine environment, and promote the well-being and welfare of poor and disadvantaged groups. The requirements for the preparation of integrated management plans for the coastal zone are many, and some cannot at present be fully met by national and local authorities in developing countries. This points to the importance of appropriate technical assistance programmes as well as financing mechanisms for the formulation and implementation of development projects.

> It has become apparent that some parts of the Convention are not being fully implemented. It must be recognized that the main value of the Convention to the community of nations resides in its comprehensive character.

However, most important of all is the political commitment to address the many problems of the coastal zone and to formulate a longer-term vision for coastal areas that is consistent with overall development goals which reflect the needs, problems and priorities of coastal populations. Programmes in this area must encompass efforts to increase awareness of the problems of the coastal zone and their implications for sustainable development and to build the support, constituencies and institutions required for implementation.

Sustainable exploitation of the resources of the EEZ

The provisions for EEZs are among the major innovations of the Law of the Sea Convention. They extend the concept of permanent sovereignty over natural resources, with EEZs serving as potentially powerful means for achieving greater equity in the oceans. However, the provisions for EEZs only acquire real significance when coastal states possess the capability to exploit and manage the resources to which they have exclusive rights. At present, this capacity is lacking in many developing countries, with the consequence that important opportunities for pursuing greater equity in the oceans remain unrealized.

● **There is a significant and growing need for programmes to support developing coastal states in the exploration, development and management of the resources of their EEZs.**

Such support should be multi-faceted, focusing on strengthening the capacity of the institutions responsible for the management of resources, including the assessment of the environmental effects of exploitation.

Many of the initiatives required in relation to the management of EEZs could most appropriately be organized at *the regional and sub-regional level*. This may help to encourage cooperation, where appropriate, in the development of management regimes for shared resources which could serve to promote cooperation in other areas, such as the joint management of freshwater resources and the control of land-based sources of marine pollution. Cooperation in these and related areas could be expected to contribute to the attainment of equity objectives in the coastal zone.

...most important of all is the political commitment to address the many problems of the coastal zone and to formulate a longer-term vision for coastal areas...

● **Regional systems for sustainable development and related marine science and technology should be established.**

Of the capacities required to exploit and manage the resources of the sea, none is more important than that pertaining to the command of science and technology. Capacities to exploit and manage the living and non-living resources of the sea are critically dependent upon the availability of technology in the form of both 'hardware' and the 'software' embodied in humans and without which the 'hardware' may be alien and unusable.

This highlights the need for new initiatives aimed at enlarging the access of coastal states to the technologies required for sustainable resource exploitation and at strengthening the capacities of developing coastal states for technological development and innovation. Such initiatives should take account of the fact that many ocean-related technologies are at present undergoing a major transformation. Much of the technology required for the exploitation of non-living resources is no longer dependent on specialized heavy equipment. The techniques now used are much more sophisticated and knowledge-intensive, and carry reduced risks for the ocean environment. Viewed from the perspective of developing coastal states, this transformation brings both constraints and opportunities. It also poses exacting requirements for technology transfer and development programmes.

In this context, it should be noted that the 'Resolution on development of national marine science, technology and ocean service infrastructures', submitted by the Group of 77, which appears in Annex VI of the Final Act of the Law of the Sea Convention, has not been implemented to any serious extent. It is more important than ever to realize the objectives of this Resolution.

A new impetus should be given to the establishment of national and regional marine scientific and technological centres by a redefinition of their objectives, character and operations. In this context, four considerations should be stressed.

● In line with contemporary developments in concepts of cooperation, involving networking and the use of the Internet, these centres should not be conceived vertically as physical entities but rather horizontally as interactive systems.

● These systems should be based on the most advanced concepts of technology development, generating a synergy between investments from the public and private sectors – at the sub-regional or regional levels – and establishing new opportunities for creative partnerships between them.

● The systems should be conceived in such a way that they not only serve the needs arising from the Law of the Sea Convention but also those from other global agreements and programmes, such as those on climate change, biodiversity and land-based sources of pollution as well as Agenda 21.

● The systems should adopt an agreed list of priority technologies – for example, aquaculture, desalination, and renewable energy from the sea (such as marine biomass) – to be derived, as far as possible, from the above agreements and programmes.

Cooperative programme for the International Seabed Authority

The International Seabed Authority (ISBA) was established under the Convention to oversee the exploitation of the non-living resources of the seabed in areas beyond national jurisdiction. At the time of the decision to establish ISBA, there was a strong expectation that deep seabed mining for manganese nodules would soon be initiated by large consortia and state enterprises. However, this development has not taken place for reasons that include the availability of land-based resources, the questionable economic viability of nodule mining, technological difficulties and large environmental problems. Deep seabed mining for manganese nodules is not now expected to take place until well into the next century. As a consequence, benefits especially to developing countries, including land-locked and geographically disadvantaged countries, have failed to emerge.

Another reason that explains the current lack in the mining of manganese nodules is the subsequent discovery of new mineral formations that may be economically more attractive than nodule mining. These have included polymetallic sulphides, methyl hydrates and, most recently, mineral deposits associated with hydrothermal vents.

Recent discoveries and their associated commercial potentials should provide a new impetus to the work of

The International Seabed Authority

The International Seabed Authority (ISBA) is an independent international organization through which States parties to the Law of the Sea Convention regulate seabed minerals exploration and exploitation, in accordance with the 1982 Convention as modified by the 1994 Agreement. The Authority came into existence, upon the entry into force of the 1982 Convention, on 16 November 1994 and became fully operational as an autonomous body in June 1996, when it took over the premises and facilities in Kingston, Jamaica, previously used by the United Nations Kingston Office for the Law of the Sea. The Authority's governing bodies are a plenary Assembly and a 36-member Council. Its subsidiary organs are the Legal and Technical Commission and the Finance Committee. An international secretariat is servicing the Authority.

The 1994 Agreement calls for an evolutionary approach to the establishment of the Authority, in keeping with its functional needs. During the period before any interest in commercial exploitation emerges, the Authority is encouraged to concentrate on, among other things: processing of applications for approval of plans of work for exploration and monitoring their implementation; monitoring and reviewing trends and developments relating to deep seabed mining activities; studying the potential impact of mineral production from the Area on the economies of developing land-based producers of those minerals which are likely to be most seriously affected; adopting rules, regulations and procedures incorporating applicable standards for the protection and preservation of the marine environment; promoting marine research, with particular emphasis on research related to the environmental impact of mining activities; and monitoring relevant technology developments, especially those relating to marine environmental protection.

Source: International Seabed Authority (1994).

ISBA and re-establish prospects for exploitation and management regimes that embody benefit sharing based upon the concept of the common heritage of mankind.

● **ISBA should prepare a long-term programme of action, for implementation at the beginning of the next century, linked to the development of 'old' and 'new' mineral resources of the seabed in ways that foster benefit sharing.**

The programme would cover the joint development of human resources and of technology, and would establish operational linkages with other bodies and programmes aimed at the testing and development of environmentally sound technologies; and at assessing their long-term impacts on cycles of material exchange, on adjacent life forms, and on the behaviour of the food web on the ocean floor.

Genetic resources of the deep seabed

New discoveries have not been confined to non-living resources. They have also included living resources, notably marine life associated with hydrothermal vents. Preliminary studies of these unique forms of marine life have indicated that they may have commercial applications, especially in the pharmaceutical industry, and the expectation is that these genetic potentials will be exploited before mineral resources.

Although generally found beyond national jurisdictions, the genetic resources of the deep seabed at present fall outside the competence of ISBA. At the time of the negotiations of the Law of the Sea Convention, almost nothing was known about the commercial potentials of the genetic resources of the deep seabed. As a consequence, the Convention makes no provision for bioprospecting, and the mandate of ISBA is confined to non-living resources.

The issues raised by these genetic resources are not unlike those raised years ago by nodule mining. These issues include the identification and evaluation of the resource potentials, the establishment of their legal status, the development of arrangements that provide for the equitable sharing of benefits from their exploitation, technology development and transfer, and the assessment of environmental impacts and implications.

• The potentials of the genetic resources of the seabed should become the subject of urgent study, focusing on their legal, environmental and economic implications, and negotiation leading to their inclusion within an appropriate international regulatory regime.

Improved standards of shipping

World seaborne trade has grown nearly six-fold in the period 1955–1995 (see Chapter 4). Traditional maritime nations have witnessed a decline in their merchant fleets, while new registers have grown rapidly. However, some of the new registers lack resources, experience and expertise. As has been observed by IMO and other international bodies, there is a need to guarantee the safety of shipping operations, the security of seafarers and ship passengers, as well as to promote clean seas. This means that continuous efforts must be made to achieve higher standards of shipping all over the world, to ban the use of sub-standard vessels and, more generally, to assist new registers in their efforts to compete in world shipping.

More attention must also be given to security on ships and to the working conditions of seafarers which, in the case of many new registers, remain below internationally agreed standards. Greater efforts are required on the part of governments to enforce the application of these regulations, through regional Port State Control organizations and other institutions. Enhanced international cooperation is also needed to assist governments of developing countries to train sailors and maritime administrators as well as to improve teaching methods in their maritime academies.

New initiatives in resource mobilization

The oceans are a source of wealth and this wealth can contribute to the promotion of equity in a variety of important ways. The introduction of even very modest levels of taxation and charges on the use of the oceans would mobilize very significant financial resources that could be used in two main ways.

First, they could serve as a general source of development financing in recognition of the fact that the oceans are a common resource and all should benefit from their use, whilst sharing in the costs of their preservation. The funds could be allocated to international programmes aimed at accelerating the social, economic and technological transformation of developing countries. Land-

As has been observed by IMO and other international bodies, there is a need to guarantee the safety of shipping operations, the security of seafarers and ship passengers, as well as to promote clean seas.

71

Translating the rhetoric of equity into practical policies that benefit disadvantaged and vulnerable people and nations is especially difficult in a global political climate that continues to be dominated by strong ideological convictions about unrestricted freedom of the seas and the importance of the international market place.

locked and geographically disadvantaged countries, as well as the least developed countries, should figure prominently as beneficiaries of the funds. The use of funds for this purpose would provide for greater automaticity in resource transfers to the developing countries, thereby placing international programmes for development cooperation on a more stable footing.

Secondly, the resources mobilized could be used to finance ocean-related development activities, either through existing mechanisms, such as the Global Environment Facility, or through the creation of new financing facilities. The use of the funds generated for such purposes would help to overcome the present shortage of ocean-related research and development activities. They could also contribute to the financing of programmes that are guided by the principles of equity and are designed to enable developing coastal states to take full advantage of their marine potentials.

Active advocacy

The pursuit of equity on both land and in the sea has always been associated with active advocacy and with the activities of formal and informal organizations that are guided by moral and ethical considerations. This will continue to be the case, and new initiatives should be considered that enable civil society to pursue issues relating to equity in the oceans.

Of particular importance are the activities of NGOs. The NGO universe is a very large one with organizations established to address a wide range of issues. Many NGOs have been formed in support of poor and disadvantaged groups in developing countries, and many exist in support of the environment, and peace and security issues.

Cooperation between NGOs active in different areas is becoming more commonplace as they seek to increase their impacts and effectiveness, and increased cooperation between NGOs in support of the oceans should be actively pursued.

Building support for the goal of increased equity in the oceans is a formidable challenge, especially as equity and the oceans have so far not been joined in a persuasive or politically acceptable manner. Translating the rhetoric of equity into practical policies that benefit disadvantaged and vulnerable people and nations is especially difficult in a global political climate that continues to be dominated by strong ideological convictions about unrestricted freedom of the seas and the importance of the international market place. But there are encouraging signs that a more positive mood is beginning to influence the political and moral imagination. For this reason, it seems justifiable to articulate equitable goals even if the means for their realization are unlikely to be forthcoming in the immediate future. The overall objective must be to encourage reflection that leads to an improved understanding of the oceans, of their immense contribution to human well-being and of the increasingly pressing problems arising from their use and abuse.

The quest for equity in the oceans thus has many dimensions. Of these, the ability to command science and technology is one of the most important. It is this command that distinguishes richer from poorer countries, with lagging capacity for technological development being one of the major cause of the growing disparities between richer and poorer nations. Nowhere are these disparities more in evidence than in the oceans. This makes ocean science and technology an issue of major importance and it is to this issue that we will now turn.

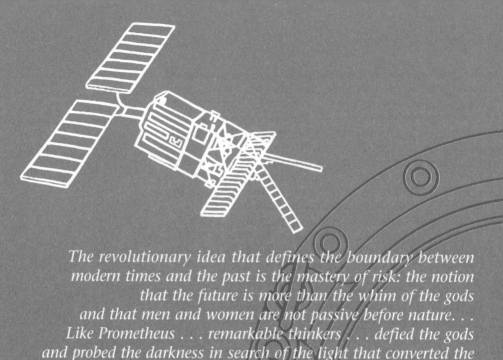

*The revolutionary idea that defines the boundary between
modern times and the past is the mastery of risk: the notion
that the future is more than the whim of the gods
and that men and women are not passive before nature. . .
Like Prometheus . . . remarkable thinkers . . . defied the gods
and probed the darkness in search of the light that converted the
future from an enemy into an opportunity.*

*Peter Bernstein, Against the Gods:
The Remarkable Story of Risk, 1997*

OCEAN SCIENCE AND TECHNOLOGY 3

The democratic and development-orientated approach to the oceans advocated in this report must be based on knowledge provided by modern science and technology. Without this knowledge, the awareness of individuals and groups will be deficient, and their participation in ocean affairs may be misdirected. Technical considerations can make it difficult for the lay person to understand and interpret the scientific phenomena underlying important public issues. Special interest groups may use their own interpretations of such phenomena to advance their own ends, sometimes at the expense of a broader public interest. In the final analysis, civil society needs to buttress itself against such manipulations, and seek to obtain the best available information about the oceans as the foundation for informed judgements.

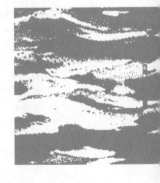

For many centuries, scientific and technological advances have driven the development of new insights into the oceans. We now have unprecedented knowledge of the oceans, and humankind's relationship with the sea, through the use of new materials, electronics, miniaturization of equipment, acoustic techniques, earth-orbiting satellites, computers and lasers. More has been learned about the nature of the oceans in the past 25 years than during all preceding history. Dramatic new insights about the seabed and life in the oceans, such as plate tectonics, hydrothermal vents and ocean–atmosphere linkages, have captured the imagination of both policy-makers and the general public. However, what we know about the oceans is still outweighed by what we do not know.

Curiosity about our planet – along with economic considerations and the availability of technology – will continue to drive ocean science, but it will increasingly be driven by a combination of growing development needs and environmental and social concerns. Such issues as pollution, overexploitation of marine living

The oceans and world climate (El Niño)

El Niño is a natural disruption of the ocean–atmosphere system in the tropical Pacific having important consequences for weather around the globe. Although it originally referred to the warm water that shows up annually off the coast of Peru around Christmas time, it has come to be a shorthand expression for what weather scientists call 'El Niño – Southern Oscillation' or ENSO. In 'normal' years trade winds blowing from east to west push surface water of the ocean heated by the tropical sun towards the western Pacific where it accumulates around Indonesia and adjacent areas. In El Niño years, for reasons scientists do not fully understand, the trade winds are less strong or may even change their directions. Warm water flows eastward across the Pacific to South America, spreading north and south and creating warmer than usual currents along the Peruvian Coast. Unusually warm ocean temperatures recorded in the central and eastern Pacific at this time are the first indication of the advent or onset of an ENSO. The ENSO increases not only the temperature of the ocean but also that of the atmosphere above the tropical Pacific.

The hot humid air makes for an initially pleasant, balmy winter in South America. But after several months there are violent, often catastrophic thunderstorms in South America, while eastern Australia and Indonesia experience severe droughts causing forest fires and sharp drops in agricultural production. The thunderstorms feed warm air and humidity more than 15 000 m (49 212 ft) into the air. The energy that is created affects high-altitude jet stream winds whose changing speeds and locations alter the weather patterns that are experienced far downstream, not only in the Americas and Asia but also as far away as Africa and even Europe. The social and economic havoc that is wrought by this interaction between ocean and wind currents can be considerable. Statistics collected from different parts of the world suggest an estimated global cost on the order of $US 13 billion for the El Niño of 1982–1983.

El Niño has never been observed so closely as the 1997–1998 phenomenon. An array of buoys stretching across the Pacific measures temperatures both at the surface and in the deep ocean. Satellites are continually collecting measurements of the height of the sea to within five centimetres. All of these data are relayed back to research centres where scientists feed them into numerical models on supercomputers in an attempt to predict the severity and incidence of the 1997–1998 El Niño in various parts of the world. It is expected to be one of the most severe of this century.

Source: US National Oceanic and Atmospheric Administration (1997).

resources, population pressures in coastal zones, harmful algal blooms, loss of marine biodiversity, the impact of environmental degradation on coastal erosion and human health can be expected to become the main drivers of scientific advance. The need to generate more energy, more food, more marine products, more wealth from the seas will also influence the course of ocean science. Global issues, such as the role of the oceans in the climate system, the El Niño phenomenon, global climate change and sea-level rise can also be expected to exert greater influence on the future direction of ocean science.

Ocean science will have to become more holistic, more interdisciplinary and more international. If we are to adequately address ocean issues at the local, national, regional and global levels, science cannot operate in isolation but will need to integrate more fully a response from society at large. There must also be changes in the way we behave towards the oceans, reflected in the way we regulate marine activities, in our social goals and our attitudes to ocean governance. If we are to make the right decisions, however, we must understand how things 'work' in the oceans and how they interact; and we must recognize the role of the oceans in our life-support system and its value for humankind. This will require excellent science, together with the technology for pursuing it, as well as the support of individuals and governments. Ultimately, it calls for a vision of the planet that embraces land, sea, the atmosphere and human societies in all their inter-actions.

It would be disingenuous to hold out the promise of scientific and technological advances to solve ocean problems, without also acknowledging that their application can itself lead to problems. The use of improved sonar equipment, fishing nets, satellite navigation, and enhanced knowledge of ocean currents have contributed to – indeed made possible – the depletion of fish stocks. Improvements in dredging equipment and seismic profiling techniques have sometimes led to the decline in, or even destruction of, demersal fish populations in the search for minerals. Our technical capacity to explore and exploit oil and gas in ever-deeper water suggests that, however deep or distant, the ocean floor may be subject to human impact.

It is not just these obvious examples of the application and development of technology that affect the oceans. Petrochemicals, artificial estrogens and many other technological innovations, apparently unconnected to the exploitation of ocean resources, have in many ways had a greater impact on the ocean's ecosystems and their living resources than ocean-based activities per se. Nor will the development of so-called 'green technologies' always have a benign influence on the oceans. For example, the development of tidal energy may require major engineering works with heavy environmental impacts in coastal areas. This is not to suggest that we must suppress scientific and technological progress, for this should be the basis on which the oceans can be used in a sustainable manner for the benefit of humankind.

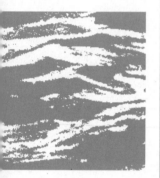

● **Science and technology must be promoted in order to realize the potential of the oceans to contribute sustainably to the satisfaction of basic needs (especially food, water and energy). We must ensure systematic prior assessment of the environmental and social impacts of new technologies.**

THE CHALLENGE OF TECHNOLOGICAL CHANGE

Some recent technical developments have related specifically to the uses of the oceans. Examples of marine advances include the design of ships, marine propulsion systems, cargo handling equipment, fishing gear, and equipment for processing at sea. Other important developments have included safer drilling rigs and spar buoys, improved acoustics and diving equipment, new corrosion-resistant materials and devices for saving human lives.

Other applications of no less importance come from more general technological advances that have greatly facilitated the use of more powerful and reliable procedures in marine science. These include robotics, semiconductors and computer modelling, electronic positioning systems and other aids to navigation, remote sensing systems, and new methods of fast geochemical analysis.

The development of technology today is almost entirely in the hands of private enterprise. The deployment of technology has been mediated in diverse ways by public authorities at different levels. Such mediation may have taken the form of restricting or even banning the use of new and possibly more efficient technologies. It may also have taken a more positive approach aimed at ensuring that technologies are more 'benign' towards the oceans, although this usually occurs as a result of public pressure after unacceptable damage has already been done to the marine environment. The use of double-hulled tankers and river sediment cleaning techniques are recent examples of this more positive approach.

The consequences of technological change can be surprising. For example, the practice of filling ships' holds with sea water as ballast is now believed to have contributed to the translocation of marine organisms. Some of these marine organisms are dangerous in their new environments, with the virtual destruction of some fisheries in the Black Sea serving as a case in point.

Such unpredictable consequences tell us that we need to: (i) know past trends in technology and to have indications of what is already 'in the pipeline'; (ii) be able to assess their potential impacts; (iii) have institutions that are more responsive to innovations; and (iv) apply the precautionary principle to the introduction of new technologies in the oceans.

An approach embodying these requirements could take two complementary forms. The first would be the designation of Marine Protected Areas (MPAs), such as coral reefs, seagrass beds, mangroves, gravel plains and other critical marine habitats. These areas would serve as havens for biodiversity, hedges against uncertainty of the consequences of technological innovation, and benchmarks against which impacts elsewhere may be gauged. This is legally possible and socially and biologically desirable in areas close to coasts. The second complementary form would be the creation of a comprehensive global network that includes the full range of categories established by the IUCN World Commission on Protected Areas. In the high seas there are more legal and practical obstacles to overcome. Nevertheless, the International Whaling Commission has prohibited indefinitely the deployment of one kind of technology – that for killing whales – anywhere in the Southern and Indian Oceans.

Major ocean features, especially some recently discovered ones, have been suggested as being in need of full, long-term protection. These include such sharply defined features as hydrothermal vents and the biota living around them, ocean trenches, abyssal plains, and tops of ocean ridges and plateaux. Some less well-defined locations, such as upwelling areas, also require special attention. MPAs can guard against the loss of precious assets, including the still poorly understood 'ecological services' (see Chapter 4) that marine systems

The consequences of technological change can be surprising. For example, the practice of filling ships' holds with sea water as ballast is now believed to have contributed to the translocation of marine organisms.

provide. Apart from considerations of international law, the identification of specific MPAs calls for the close collaboration of politicians, scientists, engineers, businessmen and civil society in general.

The assessment of new technologies prior to their deployment as part of the 'precautionary principle' would entail the establishment of credible mechanisms for multidisciplinary examination of the technology, including pilot tests to help find answers to the questions it raises, and public diffusion of the findings

THE TECHNOLOGICAL NEEDS OF SCIENCE

The era of marine science emerged with the Industrial Revolution and the arrival of steam power. In 1872–1876, the *Challenger* used steel cables rather than rope to send equipment to the seabed and to bring back water, minerals and living samples for analysis in on-board laboratories. The second half of the nineteenth and beginning of the twentieth century witnessed a series of expeditions mounted by several of the newly industrialized nations – notably France, Russia, Germany, Norway, Belgium, Sweden, Italy and the United States. Ocean science moved into progressively deeper and more hostile waters, including the Southern Ocean and the Arctic.

By the turn of the century, a sufficient number of nations were involved in this activity to require formal arrangements for international cooperation. In 1902, the International Council for the Exploration of the Sea (ICES) was created and based in Copenhagen. While one eye of ICES was focused on studies of the physics and chemistry of the sea and the general biology of the enormously diverse marine animals and plants revealed by the expeditions, the other eye was following the declining fortunes of the fisheries of the North Atlantic. Within a few years, it was clear that the old idea – that the abundance of the sea was too great to be much affected by human activities – was wrong.

The technology of marine research continued to improve through the first half of the twentieth century. The Second World War, and the Cold War years that followed it, highlighted again the strategic importance of the oceans, and gave new impetus to research and the associated enhancement of technologies for seabed exploration and for monitoring the dynamic properties of water masses. In this period – around the time of the International Indian Ocean Expedition, during the first International Geophysical Year (1957), and of the establishment of the Intergovernmental Oceanographic Commission (1960) – oceanographers fully recognized that the world's oceans formed a single dynamic entity.

The movements of water between surface and bottom, and between one region and another, are much faster than was thought a few years ago. For example, radioisotopes coming from the United Kingdom nuclear facility in West Cumbria, in spite of the 1974/92 Paris and OSPARCOM Conventions, can be shown to have reached the east coast of Canada via the Arctic Ocean. But the geography of the land masses and the topography of the seabed ensure that there is enormous diversity of localized phenomena within the unity of the world ocean, and that these smaller-scale variants are highly significant for human life. Semi-enclosed seas, such as the Mediterranean, Caribbean and Arctic, have their own enduring characteristics, as do the zones we recognize mainly by their depths and topography.

A drastic change has occurred in the geographical scale of marine research and in its continuity over time. In part, this is due to the global and dynamic nature of phenomena currently preoccupying oceanographers – such as 'red tides', El Niño, migration of marine species, and the movement of ocean currents. These entail a joining of scientific efforts from different parts of the world, and a need for repeated observations over long periods of time. In part, this is also due to the considerable increase in the costs of the technical means needed for carrying out the research. Oceanographic research today involves major projects that make use of far-flung observation stations, hundreds of ships, satellite systems, underwater vessels, radar and sonar, and computerized storage of information and modelling of ocean processes. Ocean observation with satellites since the 1960s has considerably enhanced our capability of bringing a truly global perspective to the oceans.

A new generation of ocean observation satellites carrying specific instruments (altimeters, radar, instruments to measure wind speed and wave height, ocean colour sensors, etc.) will be required. *In situ* measurements – with oceanographic vessels, automatic undersea vehicles, coupled with global observation satellites – will enhance the possibilities of ocean forecasting and will thus contribute to the further development of oceanography.

The diversity of challenges is reflected in the diversity of international institutions with mandates that include aspects of ocean science. In addition to inter-governmental bodies such as IOC, numerous specialized

A drastic change has occurred in the geographical scale of marine research and in its continuity over time. In part, this is due to the global and dynamic nature of phenomena currently preoccupying oceanographers – such as 'red tides', El Niño, migration of marine species, and the movement of ocean currents.

Intergovernmental Oceanographic Commission

Founded in 1960, the Intergovernmental Oceanographic Commission (IOC) is the competent body within the UN system committed to promoting '...marine scientific investigations and related ocean services, with a view to learning more about the nature and resources of the oceans'. In fulfilling this commitment, the IOC has focused on: developing, promoting and facilitating international oceanographic research programmes; ensuring effective planning for an operational global ocean observing system; providing international leadership for education, training and technical assistance essential to global ocean monitoring and oceanographic research; and ensuring that the collected ocean data and information are efficiently husbanded and made widely available.

IOC's programmes respond to the needs for global marine scientific investigation identified by the UN and its agencies and programmes and to the needs of its Member States stemming from the implementation of the Law of the Sea Convention and relevant UNEP decisions. IOC's programmes also respond to Member State needs for regional marine scientific investigation, and their results are communicated to relevant UN bodies.

One of IOC's major aims is the strengthening of national capabilities for marine sciences and services through world-wide partnership of its Member States and their national scientific institutions. Its work on Training, Education and Mutual Assistance in the marine sciences (TEMA) has responded to Member States' ocean research and monitoring needs connected with global warming and climate change.

IOC could be in a position to supervise and encourage regional activities so as to insure that they benefit the whole international oceanographic community and that they contribute to the attainment of common goals of international policies for ocean management. However, the lack of coordination between international bodies is an inhibition to greater effectiveness of the present international arrangements.

Source: IOC (1997).

scientific bodies exist that are concerned with the oceans, Antarctica and environmental science, many falling under the umbrella of the International Council of Scientific Unions (ICSU). Global institutions are matched by many regional bodies, each with their own mechanisms for promoting marine science, for conducting ocean observation and monitoring, and for evaluating the policy implications of the research they undertake. Prominent examples are the many regional organizations engaged in studies of fisheries resources and in regulating fishing operations. This multiplication has given rise to fragmentation and concomitant problems of inadequate communication, cooperation and coordination.

Ideas about biological productivity in the oceans are now changing rapidly, with more focus on bacteria, marine fungi and the 'nanoplankton' of exceedingly tiny but enormously abundant organisms. DNA 'fingerprints' are revising our ideas about the evolutionary relationships among marine organisms, and new techniques of observation and sampling are changing our ideas about their numbers. At the same time, the better known marine animals, such as squids and octopi, whales and dolphins, sharks and tunas, turtles and seals, are receiving new kinds of attention as we begin to obtain insights into their private lives: their sensory organs, their behaviour and their patterns of birth and death.

The other aspect of such scientific revelations is that they force us to recognize the depth of our ignorance and the pervasiveness of uncertainty. Despite the fact that, with computerized methods, more data can be collected in an hour than the *Challenger* could collect in five years, we have problems in organizing this mass of data and extracting 'natural truths' from it.

ASSESSING TECHNOLOGY FOR RESOURCE USE

Very different technologies are broadening perspectives for future supplies of minerals and energy. It is only 40 years since the first UN Conference on the Law of the Sea (1958) felt that the prospects of being able to drill holes in the seabed for oil and gas exploration and of scientific research at depths greater than 200 m (656 ft) were remote. Even when the deep seabed became of interest in the wake of Arvid Pardo's historic (1967) address to the UN General Assembly, attention was concentrated mainly on the possibility of dredging manganese nodules from the seabed. Now, submersibles

Modern fishing practices are extremely wasteful, with fishing vessels discarding unsaleable catches equal to one-third of the fish landed at markets. Of that which is used, more than one-third is converted to fishmeal for use in livestock raising and the culture of carnivorous fishes. This means that less than one-half of the marine animals caught is actually used for human consumption.

regularly survey the greatest depths. Research vessels catch marine organisms at depths of several kilometres, and there is discussion of possible ways of exploiting resources deposited in and near hydrothermal vents, the existence of which has only recently come to light. With attention currently focused more on the oceans as a future source of 'cleaner' energy supplies than of metals, news of the great abundance of methyl hydrates (substances formed at high pressure and/or low temperature as complex lattices of methane and water molecules) on the seabed at depths of at least 500 m (1640 ft) has stirred interest in circles that extend well beyond the oceanographic community.

The message that marine fisheries are, in many places, in serious biological, economic and social trouble has been diffused world-wide by FAO and other bodies, as has the parallel message that we are now close to, or perhaps past, any sustainable catch. While increases are still possible in a few places, in most others the over-whelming need is for restraint in the growth of 'fishing power' to achieve greater sustainability and to allow depleted fish stocks to recover. Modern fishing practices are extremely wasteful, with fishing vessels discarding unsaleable catches equal to one-third of the fish landed at markets. Of that which is used, more than one-third is converted to fishmeal for use in livestock raising and the culture of carnivorous fishes. This means that less than one-half of the marine animals caught is actually used for human consumption.

While the management of fishing for both high production and sustainability will always call for biological knowledge, more attention is now being given to computer modelling of the management process, including its economic and social compo-nents, rather than merely the dynamics of the fish popu-lations. Techniques of computer simulation have made this possible. Interestingly, this new approach was pio-neered by the International Whaling Commission, where a moratorium on commercial whaling gave sci-entists relief from the demands of advice about next year's catch quota. Their findings have profound impli-cations for the scientific management of fishing.

It has become clear that, if due account is to be taken of continuing and inevitable uncertainties, and inadvertent depletion of resources is to be avoided, then implementation of the precautionary principle means that fish stocks must be maintained much closer to their pre-fishing levels than was previously thought necessary, and catching must be accordingly restrained. The way to satisfactorily sustain high long-term catches is by allowing the take of a small proportion of relatively large restored stocks rather than of a large proportion of reduced stocks. But we know that fisheries of wild stocks, no matter how well they are managed, cannot feed the world. The value of fish must thus be seen more in qualitative than quantitative terms.

FAO has set the goal of doubling, by the year 2010, the part of the supply of food fish that originates from farming, or 'aquaculture'. The extent to which mariculture, or the farming of marine species, is able to contribute to the attainment of this goal will depend upon the availability and preservation of relatively unpolluted coastal waters; aquaculture will have to compete with other uses of space in coastal zones. Although a significant fraction of mariculture production consists of seaweeds and herbivorous filter-feeding molluscs – such as blue mussels, oysters and scallops – the fastest growing segment of mariculture is the cultivation of carnivorous animals. As presently conducted, this is nearly always dependent on the availability of supplementary feedstuffs derived to a considerable degree from catches of large stocks of small wild fishes. Excess of urea, combined with the use of chemicals to treat diseases endemic in crowded fish populations, has led to severe pollution problems. These are problems that must be resolved if mariculture is to thrive in the long term and to contribute significantly to human well-being. A solution might be found in the culture of more herbivorous species.

Because they have adapted to severe environmental conditions – heat, cold and high pressure – marine organisms often possess unique structures, reflected in their metabolic pathways, reproductive systems, and sensory and defence mechanisms. Many marine organisms are thus the source of biologically active compounds, having developed a kind of chemical arsenal involving molecules which signal their presence or their intention to attack prey or to deter aggressors. Some of these substances have a pharmacological interest as antibiotics, anti-inflammatory, anti-tumoural and anti-cancer agents as well as analgesics. One of the powerful new antibiotics discovered in recent years, cephalosporin, was, for example, extracted from a marine fungus. New compounds extracted from a Pacific sponge are presently undergoing clinical trials as anti-inflammatory agents, and enzymes produced by marine bacteria may have important applications due to their range of unusual properties. It is thus important to identify and assess the usefulness of bioactive compounds of marine origin so as to develop new lines of selectively active molecules with applications in the pharmaceutical and chemical industries.

The application of modern techniques of molecular biology and genetics to exploited species of fish and molluscs also represents a real challenge to fish farming. Research on gene transfer from one species to another opens the possibility of producing varieties with better and more rapid growth faculties, or greater resistance to pathologies or severe weather conditions. This would require better knowledge of the genomes of the exploited populations, of the viruses which are at the origin of the pathologies, and of the physiology of their growth and reproduction. Risks related to the introduction of gene-transfer techniques would need to be correctly assessed.

It is not only in fishing that we have to face uncertainty and risk. The safety of ships and installations, protection from natural hazards, environmental catastrophes, as well as phenomena that are less easy to quantify, such as the gradual destruction of natural habitats and marine biodiversity, make demands on science as well as on technology.

Two studies have recently been published which illustrate extraordinarily well the vast scale of ocean science, the connectivity between regional and global phenomena, and the role of the creative thinker in revealing the secrets of nature. The first of these is an analysis of records of locations and dates of catches of whales, since 1930, near the Antarctic ice edge. This has shown that the circumpolar ice edge retreated several degrees of latitude between 1954 and 1972, corresponding to a dramatic 25% reduction in the area of ice cover, which probably has significant consequences for the biological productivity of the Southern Ocean. This major regional climatic change – perhaps signalling a global one – could not have been detected by satellite observations, as they only started in the early 1970s.

The second study, providing more a 'prediction' from coupled ocean–atmosphere models than a 'discovery', entailed the development of a special computerized model to assess the likely effect of increases in global carbon-dioxide emissions on the stability of the North Atlantic circulation. It was found that a substantial increase in these emissions would bring the North Atlantic Conveyor to a halt, causing Europe's climate to cool. In this connection, it is worth recalling that studies of deep ocean sediments in the Greenland ice cap have shown that, in the past, the Conveyor system has

The circumpolar ice edge retreated several degrees of latitude between 1954 and 1972, corresponding to a dramatic 25% reduction in the area of ice cover, which probably has significant con-sequences for the biological productivity of the Southern Ocean.

stopped several times after pulses of fresh water – originating from melting snow and ice – had entered the Atlantic, and that this caused cold spells which lasted for up to hundreds of years.

Commentators on this work have noted that the proven recurrence of rapid changes in the state of the oceans makes the management of marine systems a (relatively) short-term task in which we must be prepared to manage change and, if possible, rationally adapt to inevitable change, rather than expect to maintain the *status quo*. Setting out the criteria for management in such circumstances will be an interdisciplinary task, and a challenge for both scientists and politicians.

THE NEED TO KNOW MORE – AND TO SHARE KNOWLEDGE

Coastal areas, from catchment basins to shelf edge, form a densely populated, highly complex and dynamic environment that is affected by natural processes and transformations caused by human intervention. The management of these areas requires an understanding of the physical, chemical and biological phenomena which influence their morphology, their erosion and the evolution of their ecosystems. Models to explain and forecast phenomena, such as algal blooms or the dispersion of pollutants, have to be developed and tested. Policy-makers responsible for coastal management have relied on natural scientists to identify and estimate risks to human health, living resources and the marine environment arising from economic activities. However, social scientists have yet to be involved in decision-making.

At the same time, the pace of scientific and technological progress, while greatly benefiting some societies, has marginalized or even disenfranchised others. Developing countries have lacked the capability to keep abreast of the bewildering array of new ocean knowledge in recent years, and have lacked the financial resources needed to obtain access to the required equipment. Even in industrialized countries, traditional communities and indigenous peoples have felt threatened by advances in ocean science and technology.

Given these concerns, the strengthening of scientific and technical competence of all nations is necessary in order to permit full participation in scientific and technological progress and in the benefits derived from the uses of the oceans. The effectiveness of international programmes of cooperation depends on stable institutions and qualified human resources. Moreover, scientific

We are unable to ascertain to what extent climate change will be accompanied by an increase in the number and severity of storms or hurricanes. This information is critical to coastal engineering design and to the assessment of risk by the insurance industry.

institutions must be encouraged to release and share knowledge and to participate in public awareness campaigns and educational and training programmes in other countries. One very effective way of doing this is through major ocean science cooperative programmes supported by capacity building. The range of ocean-focused topics in this category – sometimes referred to as 'mega-science' – may be illustrated by three examples: observation of ocean–climate interactions; scientific ocean drilling; and hydrothermal ocean processes and ecosystems.

The oceans and global climate

The issue of the greenhouse effect or global climate change is inextricably linked the ocean processes. The rise in sea level is an obvious example of the link and one that causes deep concern to many small island states and to countries with large populations in low-lying coastal areas. The scenarios developed by climatologists show that human-induced climate change could result in a global mean temperature rise of 1.5–4.5 °C (34.7–40 °F) accompanying sea-level rise of up to 50 cm (2 in) by the year 2100. The associated risks have to be assessed, and specific measures devised in order to protect these areas.

Our knowledge of the ocean–atmosphere system, while improving, is still limited. We do not yet fully understand the extent to which CO_2 emissions will be countered by CO_2 absorption by the oceans. Nor do we know to what extent an increase in ocean temperature will result in increased cloud formation, which in turn would decrease the earth's surface temperature. Our knowledge of the impact of major volcanic eruptions on world climate is also inadequate. We are unable to ascertain to what extent climate change will be accompanied by an increase in the number and severity of storms or hurricanes. This information is critical to coastal engineering design and to the assessment of risk by the insurance industry. Rising ocean temperature will undoubtedly affect fish stocks and migratory patterns, but by how much we do not know.

But there is much we do know and our capacity to model, to forecast, and to assess risk is rapidly improving. Whether we will be able to provide the highly sophisticated models necessary to fully understand the role of the oceans in the climate system is less certain, but the possible seriousness of the social and economic effects makes it imperative that the scientific community be enabled to 'deliver' the knowledge that society and decision-makers need.

In fact, science and technology have the potential to do more than just predict. They may even enable us to mitigate some of the negative effects by, for example, slowing down the rise in atmospheric CO_2. Experiments have been conducted to establish whether increasing oceanic primary productivity – for example, by seeding the ocean surface with iron or with fertilizers – would increase the capacity of the oceans to absorb CO_2. The answer seems to be that its impact on atmospheric CO_2 would be very limited indeed and, in any case, it is questionable whether natural marine systems should be manipulated to such an extent.

Similarly, research is presently under way on CO_2 disposal into deep ocean water. Again, there are reservations about this method in terms of its impact on bottom dwellers and on ocean chemistry as a result of rising pH. Perhaps more promising is the sequestering of liquefied or frozen CO_2 in the sediments below the ocean floor, thus theoretically avoiding potentially adverse effects on marine biodiversity. A major experiment is already under way in the North Sea to test the feasibility of this method of sequestration. However, it appears to contradict what is currently known about the costs, environmental effects and the efficacy of such options. It may also be inconsistent with international agreements which are more cautious and critical. The Convention on the Prevention of Marine Pollution by Dumping of Wastes and other Matter (London Convention), for example, bans industrial waste dumping at sea, including the dumping of CO_2, while the UN Framework Convention on Climate Change and its Kyoto Protocol do not provide for parties to dump or store CO_2 in international waters and thereby to offset their emissions.

The oceans are not only a sink for carbon from the atmosphere but also a source, and it is only on balance that they have been a net carbon absorber. A prodigious quantity of carbon has accumulated at the lower depths of the sea, where a combination of low temperature and high pressure has kept another gaseous form of carbon – methane (CH_4) – frozen in the form of methyl hydrates. Naturally occurring ocean-floor hydrates are estimated to contain twice the organic carbon resources to be found in all of the earth's recoverable reserves of coal, oil and gas.

There has been much speculation about the feasibility of producing technologies that would tap this source of hydrocarbons, especially for countries that have such deposits under their coastal jurisdictions. Perhaps more significantly, methyl hydrates are also beginning to attract the attention of climate research scientists. Very little is known about how much of an increase in the temperature of the oceans, caused by global warming, would be required to generate a significant release of this frozen gas into the atmosphere where it would constitute an important addition to greenhouse gases.

● **In the context of global warming, the importance of ocean–atmosphere interactions, the role of the oceans as a CO_2 sink and the precautionary principle, it is imperative that people and governments exploit, as a first priority, the manifold opportunities that exist for reducing carbon emissions and consider only with circumspection the potential of the oceans as a site for CO_2 disposal.**

The oceans must be monitored globally in order to improve our ability to detect and predict the effects of climate change and our ability to carry out operations with maximum efficiency and safety. One mechanism for carrying out such monitoring is the Global Ocean Observing System (GOOS). The challenge is to create favourable conditions for such a cooperative venture and to encourage scientists and nations to work together to implement this innovative proposal and to turn it into a fully operational system.

Scientific ocean drilling

The ocean floor not only contains major resources, it also reveals an important part of the history of the planet. While knowledge of this history is important in itself, it is also vital to our understanding of environmental processes and to the minimization of risk. The deep seabed reveals a record of past climate change, of waxing and waning ice sheets, of sea-level changes by 100 m (328 ft) or more. The sea level stabilized at more or less its present level about 6000 years ago. We have some competence to model sea-level change, the movement of ice sheets and the variations in climate, but the precision of modelling is inadequate as a basis for policy decisions. Indeed, at this stage we cannot even be sure that in the future, 'global cooling', due to

Global Ocean Observing System

There exists, as yet, no internationally coordinated system to observe the ocean on a global scale, define common elements of regional marine environmental problems or provide data and products on which collective national response or improvement can be built. The Global Ocean Observing System (GOOS), initiated by IOC in cooperation with WMO, UNEP and ICSU, will meet this need.

GOOS is intended to provide a scientifically based *global framework or system* for the gathering, coordination, quality control, distribution and the generation of derived products of marine and oceanographic data of common utility, as defined by the requirements of a full spectrum of user groups. GOOS will be implemented through contributions of national agencies, organizations and industries, with the assistance of national and international data management and distribution bodies.

They will be encouraged where necessary to modify and enhance their activity to comply with a coordinated GOOS plan. The objectives of GOOS are the following:

(i) To specify and detail the marine observational data needed on a continuing basis to meet the requirements of users of the oceanic environment.

(ii) To develop and implement an internationally coordinated strategy for the gathering or acquisition of these data.

(iii) To facilitate the development of these data, and encourage their application in the use and protection of the marine environment.

(iv) To facilitate means by which less-developed nations can increase their capacity to acquire and use marine data according to the GOOS framework.

(v) To coordinate the ongoing operation of GOOS and ensure its integration within wider global observation and environmental management strategies.

GOOS has been defined in terms of five 'modules' ordered according to categories of perceived user interests: climate monitoring, assessment and prediction; monitoring and assessment of marine living resources; monitoring of the coastal zone environment and its changes; assessment and prediction of the health of the ocean; and marine meteorological and oceanographic operational services. These modules are obviously interrelated and will share observations, data networks and facilities, as needed, within the one integrated system.

Source: IOC/UNESCO (1997).

natural – but as yet poorly understood – perturbations, might not overtake 'global warming'. This is not to suggest that we may wait to take action on CO_2, but to point to the risks arising from our imperfect knowledge. Improving our understanding of the past through scientific ocean drilling is one of best ways of decreasing those risks. Ocean drilling is also critical for investigating the ecology of sub-surface bacteria living in sediments, which will facilitate our understanding of how microbes influence basic terrestrial processes.

Scientific ocean drilling has a unique capacity to address a range of short- and long-term issues of great intrinsic and practical value, and the Ocean Drilling Programme (ODP) and the host ship, *Joides Resolution*, have delivered extraordinary good science. However, in a few years it will be obsolete and it will be necessary to consider options for the future.

The scale of scientific and technical expertise required for constructing and operating new drilling ships and networks of laboratories, as well as the costs, are beyond the scope of any single country. Additionally, it requires access to all the oceans, including many areas under the jurisdiction of, or adjacent to, developing countries. It is therefore, essential that both industrialized and developing countries participate in this programme that will also contribute to a better understanding of global climate change. To this end, governments will need to provide long-term support for ocean drilling while creating the necessary conditions for participation of all interested developing countries. Efforts are to be made to disseminate the results of such a programme to the world scientific community.

Hydrothermal ocean processes and ecosystems

The discovery of hydrothermal vents and associated marine life in the deep sea in 1978 and the subsequent exploration of 'black smokers' and new ecosystems in ocean ridges and elsewhere have revolutionized the thinking of biologists and geologists on many basic earth processes. New insights include: the importance of chemo-synthesis as a means of fixing energy for use by numerous forms of life; the significance of circulation and mineralization of sea water beneath the ocean floor; and how metals and other materials are deposited in crusts and chimneys. A new kingdom of life, the

The discovery
of hydrothermal
vents and
associated
marine life in
the deep sea
in 1978...
revolutionized
the thinking of
biologists and
geologists on
many basic
earth processes.

Archaea, known from hydrothermal vent communities, is providing clues about how early life on earth may have emerged. This new form of animal life, which may be abundant, depends for its survival on the warm water produced by the vents and the bacteria existing there. This has stimulated interest in the possible commercial exploitation of the Archaea as a new source of genetic material with applications in pharmacology. Similarly, the metals and minerals associated with crusts and chimneys may well have commercial value.

The number of sites explored so far is a small fraction of what appears to be a widespread and extremely important phenomenon geologically, biologically, and possibly economically. To further investigate deep-sea ridges and other relevant sites will require manned submersibles and other underwater systems and sensors as well as cooperation among scientists of many disciplines from industrialized and developing countries. The vision for HOPE is that a major international effort will be organized to extend the successes already realized through cooperative projects so that the scientific communities in all interested countries, especially developing countries, can participate in this fascinating new area of ocean science.

BRIDGING THE GAPS

Ocean management decisions with economic and social implications must draw upon well-documented scientific and technical information, with decision-makers dependent on scientists for collecting and interpreting the relevant data. To this end, marine research has to be organized on a quasi-permanent basis – both nationally, regionally, and globally – in order to draw upon scientific expertise wherever it exists, while ensuring the widest possible access to data and research results. Electronic networking, data banks, and the use of information technology, all facilitate such access and hence enlarge capacities for improved ocean management.

Mining deep seabed resources

Miners exploring in a 5200 sq. km (2000 sq. mi) area of the deep territorial waters of Papua New Guinea recently laid claim to gold, silver and copper deposited at the sites of volcanic springs a mile down from the sea's surface.

Because of the richness of these deposits, experts argue that less processing will be required on land to separate out the different metals and turn them into ingots. The deposits are much richer in precious metals and closer to the sea's surface than the icy manganese nodule that litter the global seabed. This makes them much easier and less costly to recover. Although many such hot deposits of metals have been mapped out since the 1980s, no one has yet mined the rocky outcroppings which can grow many stories high. If the deep hot springs in the Papua New Guinea area turn out to be as rich and widespread as surveys indicate, miners will take out preliminary hauls of 9700 tonnes each in the next two years and large commercial loads in the next five years. Metals in this deposit could be worth hundreds of millions of dollars in the immediate future with the prospect that mining companies could turn to similar deposits in other parts of the world, in much the same way that petroleum companies have increasingly moved into deep waters.

Unlike miners, ecologists view the exploitation of the sea's dark recesses as an assault on a poorly explored region of natural wonders such as blind shrimp, giant tube worms and other unfamiliar creatures that thrive in densities that rival the life in rain forests. Since the deep volcanic hot springs are also important in evolutionary studies and are increasingly seen as the possible birthplace of life on earth, great care should be taken to protect a sufficient number of them to permit scientists to obtain an insight into how they were before the mining began.

Source: International Herald Tribune, 22 December 1997.

• Closer interfaces should be established between the marine and social sciences in support of a holistic understanding of ocean problems. Dialogue must also be strengthened between experts, decision-makers and the public, in order to gain support for more effective and integrated management – including protection and conservation – of the oceans and coastal zones. Science must form an integral part of the decision-making process.

• Although ocean science is international by its very nature, there still exist great gaps between nations and regions of the world as far as scientific and technical capacity is concerned. A major effort must be made to bridge these gaps through inter-national cooperation, bearing in mind that the geography of ocean-related problems calls for local, national and regional solutions in a global context.

Without science and technology, many of the resources of the oceans remain beyond human reach. The application of science and technology to the seas has transformed the relationship, for better and worse, between humankind and a vast domain that covers nearly three-quarters of the planet's surface. This transformed relationship is compelling us to reconsider the value of the oceans. Issues in valuation form the subject of the following chapter.

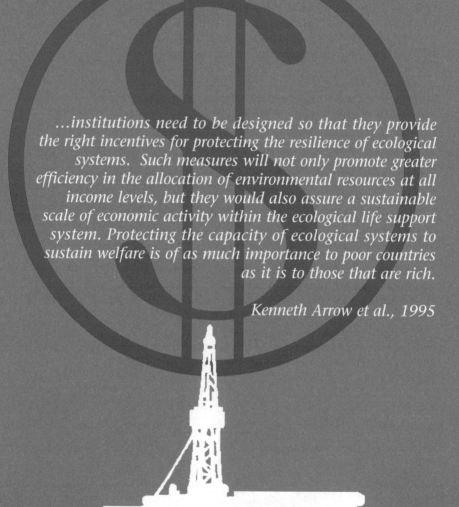

...institutions need to be designed so that they provide the right incentives for protecting the resilience of ecological systems. Such measures will not only promote greater efficiency in the allocation of environmental resources at all income levels, but they would also assure a sustainable scale of economic activity within the ecological life support system. Protecting the capacity of ecological systems to sustain welfare is of as much importance to poor countries as it is to those that are rich.

Kenneth Arrow et al., 1995

VALUING THE OCEANS 4

T he economically and ecologically sustainable use of ocean resources involves much more than improving management in individual sectors, such as fisheries, marine transport, and off-shore extraction of oil, gas and other minerals. It also involves recognition of the ways in which land-based activities affect the oceans. The enormous increase in economic activity and the settling of more and more people in coastal zones are threatening the ecological value of the oceans. The words 'economy' and 'ecology' both have their roots in the Greek word *oikos*, meaning 'our common home'. This chapter takes *oikos* as its starting point. Its main purpose is to highlight the interfaces between economy and ecology and to explore the value of the marketed and neglected, *non-marketed* services yielded by the oceans, with the intention of heightening awareness of their true contribution to individual and collective well-being.

THE PROBLEM

As world population, world economy and world trade have grown, so has the demand for marine and coastal resources. As technologies have developed, the range of resources that can be exploited has widened. The problem now confronting us is that ocean ecosystems are often used in ways that are unsustainable, not only in environmental but also in economic and social terms. All too often, the costs of this bear most heavily on the poor. Management regimes to control land-based sources of pollution, overfishing, and other threats to the stability of marine and coastal ecosystems are often poorly developed or do not exist. Access to many ocean resources – defined here to encompass both the resources themselves, including 'environmental' resources, and their uses – is unregulated and, where such regulation does exist, it is often ineffective.

People using these resources are able to ignore the costs they impose on others. Such costs may be localized and the effects of relatively short duration, as is the case with some oil spills. They may even be known with some certainty. But they may also be wide-

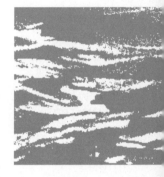

97

spread, highly uncertain, and have long-lasting or irreversible effects, such as the fundamental change that has taken place in the ecology of semi-enclosed seas as a result of land-based pollution. Liberalization may have yielded many economic benefits, but it has also increased the risk of environmental damage due to the failure of markets to register the environmental costs of economic activity. Markets have typically failed to signal the true scarcity of resources, both on- and off-shore, and liberalization has often weakened corrective regulations or incentives. Indeed, the world-wide structure of property rights, taxes and subsidies has encouraged over-use of coastal and marine resources. This has often placed these resources under intolerable stress.

An estimated 70% of the world's fish stocks are already being exploited at or beyond sustainable limits, but fishing generally continues unabated despite extensive regulatory arrangements for their management. In a few cases, emergency measures have been taken to defend stocks of marine wildlife. In 1985, for instance, the International Whaling Commission declared a moratorium on commercial whaling. Individual fisheries have similarly been closed for prolonged periods, but the threats to fisheries remain and continue to increase.

The pressure on the oceans is not only due to over-harvesting but also to the cumulative impact of land-based activities. This includes many of the effects of coastal development, especially the destruction of wetlands, mangroves and coral reefs, sedimentation and the dredging of sediments, damage to watersheds and the impounding of water supplies to support urban development in coastal areas.

The oceans have also become the ultimate sink for discharges of waste of all sorts – carried by rivers and winds – from land-based sources, including coastal megacities. Other threats come from the transport of hazardous wastes, operational and accidental spillage of oil, discharge of radioactive materials at sea, nuclear testing, and the transport of alien species in the ballast water of ships. Harmful algal blooms, nourished especially by sewage and agricultural run-off, are becoming increasingly common, adversely affecting the recreational values of many coastal areas, and in some cases reducing fish populations and producing anoxia in the water column.

An estimated 70% of the world's fish stocks are already being exploited at or beyond sustainable limits ...

While inputs of some toxic pollutants – such as halogenated hydrocarbons and polychlorinated biphenyls – have declined in the North, they continue to rise in the South. More importantly, they continue to accumulate in all the world's coastal zones. As much as 90% of all waste material entering coastal waters remains there in sediments, wetlands, mangroves and coral reefs. The cocktail of fertilizers, pesticides, sewage effluent and industrial discharges flowing down the Mississippi is believed to be responsible for the 'dead zone' that lasts approximately eight months a year and extends over thousands of square kilometres in the Gulf of Mexico.

SUSTAINABLE USE OF COASTAL AND MARINE RESOURCES

If the development of coastal and marine resources is to be sustainable, the benefits the resources yield at present should not impair their capacity to benefit future generations; users should, therefore, take all the effects of their actions into account. The sustainable development of coastal and marine resources is potentially threatened by a combination of factors, including the growth in demand for ocean resources, the failure of markets and policy, and poverty. Growth in demand stems both from population growth and the expansion of economic activity, especially where such activity has aimed not only at satisfying the growing demand for existing products but also at generating a steady stream of new products that cater to affluent consumers.

Market failure implies that the market prices that guide individual decisions to use resources often fail to reflect the damage that may result. At the same time, governments may neither provide users with incentives to conserve resources nor protect those at risk when markets fail.

The value of coastal and marine ecosystems

The value of ocean resources stems from the various goods and services they provide. In the area of oil and gas, off-shore production already accounted in 1995 for 26% of the world's total. The contribution to biological diversity can be appreciated by the fact that 15 of the 33 types of animal life or 'phyla' on the planet are found only in the oceans. Marine fisheries account for 85% of the global fish catch. Maritime shipping is involved in the transport of over 80% of the world's merchandise trade.

Commission on Sustainable Development

The Commission on Sustainable Development (CSD) was established, under ECOSOC, to review the progress in the implementation of Agenda 21. In 1998–2002, the CSD will be implementing its new work programme, adopted by the June 1997 Special Session of the General Assembly (Earth Summit+5). This continues to emphasize the important role of major groups in sustainable development and the need to assure that their contributions are included.

Throughout the 1998–2002 programme, poverty and changing consumption and production patterns are overriding issues.

Year   Themes

1998   **Freshwater Management**
        Transfer of Technology/Capacity-building/
        Education/Science
        Industry
        Sustainable Development of Small Island Developing
        States (SIDS)

1999   **Oceans and Seas**
        Consumption and Production Patterns
        Tourism
        Comprehensive review of the Programme of Action for
        the Sustainable Development of SIDS.

2000   **Land Resources**
        Financial Resources/Trade and Investment/
        Economic Growth
        Agriculture
        Day of Indigenous People

2001   **Atmosphere/Information for Decision-Making and**
        Participation
        International Cooperation for an Enabling
        Environment
        Energy/Transport

2002   Comprehensive review Agenda 21 implementation
        (Earth Summit +10)

During this cycle of five years, the Division for Sustainable Development is considering a number of special events to highlight specific major group sectors or the role of major groups in general in relation to the thematic issues.

Source: Commission on Sustainable Development (1997).

Marine transport and ports
(Goods in million metric tonnes)

Growth of world seaborne trade

	Trade	**Per head of world population** (in metric tonnes)
1955	800	0.286
1975	3064	0.747
1995	4700	0.832
2000 (est)	5690	0.940

Ten largest ports in the world (latest year available)

1.	Singapore	306	(1995)
2.	Rotterdam	294	(1995)
3.	Chiba	193	(1992)
4.	South Lousiana	172	(1996)
5.	Houston	134	(1996)
6.	Hong Kong	127	(1995)
7.	Nagoya	124	(1995)
8.	New York/New Jersey	119	(1996)
9.	Antwerp	108	(1995)
10.	Yokohama	108	(1995)

Source: Gary Crook, UNCTAD Secretariat (1998).

Offshore oil and gas production as % of world total 1993 – 1995

The shares of offshore oil and gas production of the world total production are given below for the period 1993 – 1995. (mtoe = million tonnes of oil equivalent)

	1993	**1994**	**1995**
Oil production			
- offshore (mtoe)	880.9	932.7	965.3
- world (mtoe)	3182.5	3224.3	3265.4
- share offshore (%)	27.7	28.9	29.6
Gas production			
- offshore (mtoe)	364.3	381.1	397.9
- world (mtoe)	1860.8	1881.3	1915.1
- share offshore (%)	19.6	20.3	20.8
Oil + Gas production			
- offshore (mtoe)	1245.2	1313.8	1363.2
- world (mtoe)	5043.3	5105.6	5181.0
- share offshore (%)	24.7	25.7	26.3

The shares of offshore oil and gas have steadily grown over 1993 – 1995 to around 30% and 21% of the world total production. As a result the shares of offshore oil + gas production have grown over the same period to around 26%.

Source: Offshore, April 1995 and BP Statistical Review of World Energy 1997.

Many attempts have been made to arrive at figures for the total value of ocean-related goods and services in monetary terms. These attempts have been fraught with enormous difficulties because of a lack of data and intractable methodological problems. Nevertheless, the estimates help to give at least a crude indication of the relative importance of the oceans in economic terms. One recent study suggests that the sum total of marine industries (such as oil and gas, tourism, seaborne trade, naval defence, shipbuilding, fishing, non-fuel minerals, submarine telecommunications), for which data are available, amounts to approximately US$1 trillion out of a total global GDP of US$23 trillion. While this 4% share of global GDP may appear small, the estimate takes no account of the value of 'ecological services'. A number of economists have sought to estimate the value of such services, and an overview of recent estimates is given in Table 1.

These estimates support two conclusions. First, the total estimated value of the world's ecological services, such as gas regulation (e.g. carbon dioxide/oxygen balance and ozone for ultraviolet radiation protection), disturbance regulation (e.g. storm protection and flood control), waste treatment and nutrient cycling, amounted to around US$33 trillion in 1994. Secondly, marine systems contributed nearly two-thirds of the total, and more than half of this came from coastal systems. Coastal marine environments and wetlands (tidal marshes/mangroves) are of disproportionately high value to humankind. While they account for 6% of the earth's surface, they form 43% of the estimated value of the world's ecosystem services. Much of this value arises from their role in regulating the cycling of nutrients which control the productivity of plants on land and in the sea. Hence the importance that the framers of the Law of the Sea Convention attached to the protection and preservation of the marine environment in the EEZs, in which these coastal marine environments are located.

Where extraction or environmental degradation impairs an ecological function, the resulting loss of output gives a measure of the social cost involved. Mangroves and watersheds, for instance, protect fisheries and coastal areas. Vietnam regularly experiences major floodings of coastal and estuarine systems. However, the severity of floods in the Red River and Mekong deltas is directly related to deforestation in the relevant watersheds. The Vietnamese Disaster Management Unit estimates that the 1996 floods – which caused more than 1000 deaths and submerged some

Table 1: Estimates of the global value of annual ecosystem services in 1994

Biome	Area (million ha)	Total value per ha ($US/ha/yr)	Total global flow value ($US billion/yr)
Marine	**36 302**	577	**20 949**
Open Ocean	**33 200**	252	**8 381**
Coastal	**3 102**	4 052	**12 568**
Estuaries	180	22 832	4 110
Seagrass/Algae Beds	200	19 004	3 801
Coral Reefs	62	6 075	375
Shelf	2 660	1 610	4 283
Terrestrial	**15 323**	804	**12 319**
Forest	**4 855**	969	**4 706**
Tropical	1 900	2 007	3 813
Temperate/Boreal	2 955	302	894
Grass/Rangelands	**3 898**	232	**906**
Wetlands	**330**	14 785	**4 879**
Tidal Marsh/Mangroves	165	9 990	1 648
Swamps/Floodplains	165	19 580	3 231
Lakes/Rivers	**200**	8 498	**1 700**
Desert	**1 925**		
Tundra	**743**		
Ice/Rock	**1 640**		
Cropland	**1 400**	92	**128**
Urban	**332**		
Total	**51 625**		**33 268**

Source: Costanza *et al.* (1997).

840 000 sq. km (324 324 mi) of rice fields – resulted in losses amounting to US$655 million. These costs are not taken into account by farmers in the watersheds. Nor are they captured in conventional economic measures of well-being.

World-wide, the value of ocean resources ignored in market transactions is very high, and the potential cost to humanity when markets for coastal and marine resources fail is significant enough to warrant serious international concern.

Market prices do not give the right signals

The economic problem of coastal and marine resources stems from the fact that market prices – the most important economic measure of scarcity – are poor indicators of the value of ocean resources or of economic opportunities lost because of the way the oceans and ocean resources are used or abused. For example, 'healthy' coral reefs are characterized by high levels of fish and other aquatic diversity. If seriously disturbed, they may change into a state dominated by blue-green algae, with low levels of aquatic diversity. In this case, the economic value of the ecological system depends on its state, but the price of fish and recreational services will not signal an impending change of this state.

More generally, a change in the composition of species will simultaneously change both the ecology and the economics of the system. The most sensitive components of food webs, energy flows and bio-geochemical cycles are those where the number of species carrying out key functions is very small. There are limits to the depletion of species or pollution of such systems beyond which they lose resilience and are unable to deliver ecologically and economically valuable goods and services. However, market prices are unable to signal changes in the risk that limits are being exceeded. The management problem is how to ensure that the institutions governing the allocation of coastal and marine resources, and incentives and disincentives to resource users, protect the resilience of coastal and marine ecosystems.

Use rights to safeguard sustainability

The failure of prices to signal the actual scarcity of resources is generally ascribed to the structure of property rights. In coastal and marine systems, the present structure of rights is complex. The ocean is evolving from a resource with open access – *mare liberum* – to a

World-wide, the value of ocean resources ignored in market transactions is very high, and the potential cost to humanity when markets for coastal and marine resources fail is significant enough to warrant serious international concern.

common resource that is subject to a mixture of private and regulated access. The ability of markets to signal resource scarcity tends to be weak where resource access is open.

Under the Law of the Sea Convention, coastal states have an extended national jurisdiction in which they are able to regulate access to marine resources. In these areas, the most common form of use rights are extraction/ harvesting licences, principally for minerals and marine fauna, and non-extractive licences to use assets, such as beaches and harbours. But they also include much older common harvesting rights. Beyond the limits of national jurisdiction, access to living resources is, in principle, free but the rights of users are regulated through an expanding set of collective management agreements. Access to the mineral resources of the deep seabed is regulated by the International Seabed Authority. In both national and international waters, regulated access implies the establishment of use rights. These are 'rights' to use specific resources in particular ways; increasingly, they also include responsibilities of users.

External effects

Open access to the oceans has traditionally meant that users have enjoyed the right to dispose of waste at sea. While much of this waste has been either hydrocarbons or organic waste with a relatively short-term impact, it has also included heavy metals and nuclear waste with the potential to affect marine ecosystems for thousands of years. Indeed, until recently, the open ocean was the preferred site for the dumping of nuclear waste.

Most of the ecological effects of waste disposal, spillages and industrial and economic activity at, or on, the sea are 'external' to the resource users. That is, the cost of these activities is not paid by users and does not affect their business decisions, whereas the individuals and communities who do bear the cost have no say in the decisions that give rise to it.

The external effects of mariculture serve as a case in point. The most widespread forms of mariculture (see Chapter 3) are the cultivation of shrimp and prawn in tropical countries and the culture of salmon in temperate regions. Both can have considerable impacts on coastal fisheries – such as the transferral of diseases from cultivated to wild species, loss of habitat for wild fish, and increased ocean pollution – but such external costs were not considered when the original decision was taken to begin mariculture.

For many marine resources, it is impossible, or at least very costly, to control access. This may be due to historical rights of open access – the high seas being the classic example. Common international resources, such as pelagic fisheries or whales, have traditionally been treated in the same way. In other cases, the nature of the resource is such that access is open to all. The moderating effect of the oceans on the atmosphere, and the genetic information contained in the marine bio-diversity, serve as examples. No one can be excluded from the benefits of their conservation, but society's benefit from one person's conservation of resources is greater than the benefits that accrue to the individual. Hence, from society's standpoint, too little effort will be devoted to their conservation.

The solution to this problem lies in collective action: either to regulate access to such resources or to invest in their management and conservation. At both national and international levels, the likelihood of securing cooperation in collective action depends on the perceived value of the benefits gained relative to the costs of cooperation, and on how the benefits will be distributed among individual users of the resources, nation states and the international community. Cooperation to serve sustainability has therefore, to begin at home, building up to the national, regional and, ultimately, global levels. However, given the different levels of economic development, the industrialized countries have a special responsibility to address global problems by accepting restraints on certain economic activities in order to reduce the threat to nature, including climate.

ECONOMIC INCENTIVES IN OCEAN MANAGEMENT

If appropriate institutions, policies and instruments are to be developed to deal with the sustainability of the oceans, there must be a clear understanding of the incentives and disincentives associated with the system of property rights, market structures, tax and subsidy regimes, patterns of public expenditure, and so on. The scope for national regulation of access to resources varies between areas within national jurisdiction and beyond, and this alters the incentive effects of each. But the economic problem of the oceans does not only involve the incentive effects of open access or imperfectly regulated use. It also involves the on-shore incentives behind such activities as changes in land use, alterations in watersheds and coastal (especially mangrove) forests, disposal of agricultural, domestic and industrial wastes into rivers, and harbour congestion.

Policy options

To bring about an efficient allocation of environmental resources, it is important to charge users for the full social costs of their decisions. For activities within national jurisdiction, levying appropriate charges is comparatively straightforward. A number of market-based or price-like instruments may be applied. Table 2 summarizes the options that are relevant to the protection of the oceans, coastal areas and watersheds. These include taxes and tax exemptions, user fees or charges, administered prices, royalties and fines. Each has the effect of changing the private cost of a resource to the user, and if the change brings private and social costs into line with one another, the resulting decision will be economically efficient; that is, the instrument will 'internalize' the external environmental effect. In this respect, subsidies – that have the perverse effect of lowering the private cost of using coastal and marine resources – should be identified and phased out.

The precautionary principle

'Where there are threats of serious or irreversible damage, lack of full scientific certainty shall not be used as a reason for postponing cost-effective measures to prevent environmental degradation'. This is the 'precautionary approach' affirmed in Principle 15 of the 1992 Rio Declaration. One of its earliest applications was in the moratorium on commercial whaling imposed by the International Whaling Commission in 1985. It was also enunciated in the 1989 Declaration of the Third Ministerial Conference on the North Sea which called for 'action to avoid potentially damaging impacts of substances that are persistent, toxic and liable to bio-accumulate even where there is no scientific evidence to prove a causal link between effects and emissions'. Since then, the concept has been further elaborated and generally accepted, as illustrated by its introduction in a number of agreements, such as the Climate Change Convention and that on Straddling Fish Stocks and Highly Migratory Fish Stocks.

The precautionary principle is thus a statement about both acceptable levels of risk and the onus of proof. Restating the principle as defined by the North Sea Ministerial Conference, it asserts that action will be taken to avoid serious potential damage from emissions to the North Sea unless there is 'proof' that emissions will not result in such damage.

There is a link between the precautionary principle and the widespread practice of discounting. The use of

'Where there are threats of serious or irreversible damage, lack of full scientific certainty shall not be used as a reason for postponing cost-effective measures to prevent environmental degradation'. This is the 'precautionary approach' affirmed in Principle 15 of the 1992 Rio Declaration.

Table 2. Instruments for the preservation of coastal and marine systems

	Protected areas and access restrictions	Property rights	Taxes and charges	Penalties	Financial incentives	Legal liability	Bonds and deposit-refund systems
Ocean protection	Marine reserves; gear restrictions; quotas; close seasons; catch allowances; harvest moratoria	Fishing licences; individual transferable quotas	Access charges; boat licensing fees	Fines; gear and boat seizures for breach of fishing regulations	Investment incentives; depreciation allowances; tax exemptions	International conventions on liability and compensation funds	Oil spill bonds
Coastal protection	Coastal reserves; land use restrictions	Concessions for logging and forest management communal rights	Taxes; royalties; coastal defence levies	Fines and condemnation for zoning infringements	Reforestation incentives	Natural resource liability	Reforestation bonds; forest management bonds
Water pollution	Discharge consents; water quality regulations; extraction licences	Effluent licences; tradable effluent permits or discharge consents	Effluent charges; water treatment fees	Fines for illegal dumping and for breach of hazardous and toxic material regulations	Technology loans; depreciation allowances	Legal liability; liability insurance	Environmental bonds

Source: Adapted from Panayotou (1994).

a discount rate to discriminate between different moments in time means that far-future but potentially very severe environmental effects may be largely ignored in any calculation of the costs and benefits of economic activities. The precautionary principle is often invoked where the environmental effects of an activity are thought to be particularly serious, but where the discounted cost of those effects is insignificant. In other words, it is invoked to avoid decisions that society may later have cause to regret.

Precautionary instruments

Precautionary instruments are typically designed to manage ecological systems that are neither fully observable nor fully controllable, as is the case with many marine systems. The incentive effects and the cost-effectiveness of different precautionary instruments vary. The conservation measures adopted by the 1980 Canberra Convention on the Conservation of Antarctic Marine Living Resources, for example, embody a precautionary principle. They are designed to ensure that the growth of new and exploratory fisheries in the convention area develops at a rate consistent with the growth of the data needed to determine sustainable catch levels. The precautionary instrument in this case is a set of constraints on catch levels. The constraints are reviewed as data accumulate on the impact of different rates of harvest on both the target and associated species.

The insensitivity of prices to change in resource stocks or in their properties in many systems is evidence that those changes cannot be observed. In fisheries, overharvesting has seldom been reflected in rising prices; price movements have had little role to play in the conservation of fish stocks anywhere. Hence, while there may be some advantage in correcting for policy-induced distortions in the price regime, conservation of fisheries requires something more.

Precautionary instruments, such as fish harvesting quotas, protect the resilience of the system being exploited by restricting the level or the nature of economic activity affecting it. Such management instruments are commonly described as 'safe minimum standards' or 'sustainability constraints'.

Harvesting quotas or closed seasons are historically well-established examples of species-specific standards, and remain the most widely used mechanisms for the conservation of fisheries. More recently, individual transferable quotas (ITQs) have been introduced in

...the capacity
to adapt,
as information
on the state of
the system
changes, is
a critically
important
aspect of
management
regimes
based on
precautionary
instruments.

some fisheries. While some hold the view that ITQs offer potential gains in economic efficiency because the quotas are tradable (transferable), they still rely on the same basic protective mechanism as other forms of quotas, e.g. a restriction on total allowable catch.

Establishing and managing fishing quotas are problematic for five main reasons:

● Quotas do not necessarily preserve equity, especially in situations where small artisanal fishers exist side-by-side with large, financially powerful fishing enterprises. Care must therefore, be taken to ensure that certain social objectives, such as community development, are fully taken into account in the actual design of ITQ schemes. For example, proceeds from the sale of fishing rights should benefit the affected communities.

● If quota are to be set properly, precise data on the fish biomass are required, and these are never known with certainty.

● Total allowable catches are often seen as fixed, firm and sacrosanct within the fishing season, a perception incompatible with the need for mid-season quota adjustments.

● Quota management creates anti-conservationist 'perverse incentives' to harvest more fish than is allowed by the established quotas. This is frequently achieved by 'high-grading' to maximize the value of what is reported as caught and by dumping prohibited fish – for which the quotas have been reached – so as to be able to continue fishing for other stocks.

● Catch data tend to be biased both by misreporting and anti-conservationist behaviour.

It follows that the capacity to adapt, as information on the state of the system changes, is a critically important aspect of management regimes based on precautionary instruments.

INTERNATIONAL OCEAN MANAGEMENT AGREEMENTS

The difficulty of protecting the integrity of stocks in international waters is even greater than in waters under national jurisdiction. The most effective means of insuring against the overexploitation of environmental public goods, including ocean resources and ocean space, is to provide for some degree of exclusion, for instance, by restricting access to the resource to a recognized and limited number of entities. But such exclusion is, of course, only justified in so far as it is necessary to achieve sustainability.

In successfully regulating common resources, the following criteria are particularly important: clearly defined membership and responsibilities; rules that are consistent with local conditions; membership participation in decisions; effective monitoring; sanctions against the violation of rules; mechanisms for the resolution of conflicts; and external recognition. Conditions that favour cooperation in institutions of this kind include: repeated interaction between members; information about members' behaviour over time; and a small membership.

The most effective international agreements appear to: involve a limited number of signatories; have evolved through repeated re-negotiation; and include effective disincentives to defect from a precisely defined set of objectives. A good example of the latter is the 1978 Protocol to the 1973 International Convention for the Prevention of Pollution from Ships. This Protocol required that new ships be fitted with segregated ballast tanks and crude oil washing facilities. Compliance has been virtually complete because it is easily verified, and parties to the agreement may deny entry by vessels not in compliance. Moreover, tanker owners who invested in the more expensive equipment were given assurance that other owners would not be allowed to gain a competitive advantage by failing to do so.

The benefits of international agreements

The effectiveness of agreements that satisfy the above criteria depends on the distribution of benefits among the signatories. The benefits to individual nations, under an international agreement dealing with a public good, will depend on a number of considerations of which the following three are most relevant:

● Problems affecting some public goods can be solved by action of the country with the greatest capacity in that field. In these cases, the most effective contribution determines the level of the public good enjoyed by all. A good example is the role of the US Navy in making available, to the rest of the world, timely information on a variety of ocean parameters.

● There is a class of public goods in which the level enjoyed by all is simply the sum of national contributions. The conservation of stocks through harvest cuts is an example of this kind of public good. The benefit enjoyed by all depends on the total reduction in harvest.

● There is a class of public goods in which the level enjoyed by all nations is given by the least effective contribution – the 'weakest link' problem. For example, where control over a communicable disease involves eradication campaigns in all nations, control will only be as good as the campaign run by the least effective nation. Conventions need to be both stringent and binding on the behaviour of the weakest link in an agreement.

Whereas 'free-riding' in the first case has no impact on the level of the public good, 'free-riding' in the remaining cases reduces the benefits to all. Indeed, this is the classic public goods problem. Most international environmental conservation involves public good situations of the second or third kind.

Whether agreements to provide the public good are sustainable or not depends on the distribution of costs and benefits of compliance. Hence, the design of agreements is critical; so is the funding of public good provision.

Although multilateral agreements may be difficult to sustain, they are generally more sustainable than unilateral action. One reason for this is the increasingly limited room for manoeuvre available to individual countries under the multilateral trading system. A country's unilateral action to protect the marine environment through taxes, user charges and other incentive measures can, by raising costs for particular industries, reduce their international competitiveness. Industries which resist or lobby against such measures for reasons of international competitiveness are far more likely to acquiesce to them if they are part of multilateral agreements that create a 'level playing field' for all contenders.

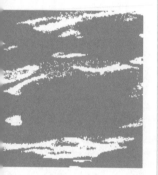

FUTURE DIRECTIONS

The unsustainable use of ocean ecosystems is reflected in progressive overexploitation of living resources, coastal and marine pollution, and disruption of marine habitats. Reversing these trends requires institutions for marine resource management that are themselves sustainable. This in turn implies that such institutions deliver positive net benefits to all stakeholders – those who derive their livelihood from the oceans and those who consume ocean products alike.

The management of the oceans must come to grips with their unique characteristics. Since the oceans are a fluid medium that recognize no boundaries, it may, depending on the circumstances, be appropriate to deal with management issues either at the community, national, regional or global level. Moreover, the types of human activities that impinge upon the oceans may have origins on land. All of these aspects should be considered in drawing up management regimes for the oceans.

A precondition for effective regional and global management regimes is, of course, that the states concerned should be able to participate, a requirement which is seldom met in the majority of developing countries. This calls for revitalized international cooperation in support of capacity building as well as for the possibility of accessing regional cooperation schemes and GEF facilities – important issues which have been addressed in Chapter 2.

A precondition for effective regional and global management regimes is, of course, that the states concerned should be able to participate, a requirement which is seldom met in the majority of developing countries.

● **Management regimes for coastal and marine resources should be established at the appropriate geographical/political level, provide for cooperation between disciplines and recognize the importance of land-based impacts.**

Gearing management regimes to the appropriate geographical level

A key element in institutional and managerial sustainability is the principle which holds that the use of environmental resources should be managed at the appropriate level. If the geographical spread of environmental effects includes several countries, all should be part of the management regime or governance structure. Conversely, if the environmental effects are purely local, then the governance framework should be purely local. Safeguarding the sustainable use of the

113

oceans begins at the community level, but economic interactions call for integrated approaches at the national level.

At the national level, there are numerous instances of regimes for the sustainable management of local resources. There are many more – both national and international – that are unsustainable. Some have presided over the collapse of the ocean resources they were set up to manage, or have caused irreversible collateral damage to linked species or systems. Others have preserved the financial viability of resource use only through massive subsidies or by shifting on to others a part of the costs of resource use. Still others have failed because the institutional arrangements or regulatory regimes were ineffective.

Stimulating cooperation between disciplines

A major problem usually resides in the failure to accommodate all relevant effects within the management regime. In some cases, this stems from a failure to appreciate the interconnectedness of geographically distant events. In others, it stems from a failure to understand the linkages between interdependent economic, social and ecological processes. In still others, it stems from a set of institutional arrangements and property rights that allows resource users to ignore the environmental effects of their actions.

If management of coastal and marine resources is to cope with the complexities of the system, it must provide for the participation of scientists from many disciplines spanning both natural and social sciences.

Managing on-shore and off-shore linkages

If ocean resources are to be both maintained and developed, it is clearly important to strengthen our understanding of the structure and dynamics of interconnected social, economic and ecological processes – and that this understanding underpins the development of management strategies.

Without an understanding of the linkages between on-shore depletion or pollution of environmental resources and changes in marine ecosystems, it will be impossible to develop sustainable strategies or markets in which prices reflect the true costs of resource use. Chapter 17 of Agenda 21, dealing with the oceans, asserts that the sustainable use of coastal and marine resources requires fundamentally new approaches at national, regional and global levels. These approaches should be integrated, precautionary and anticipatory. They should also address the range of development and environmental issues posed by the exploitation of ocean resources. These include: sustainable use of coastal and EEZ resources and of living resources of the high seas; addressing the key uncertainties associated with the evolution of marine and climatic processes; and strengthening international communication, cooperation and coordination.

This requires an understanding of the driving forces behind activities that affect ocean resources, and of the interactions between economic, social and ecological processes in coastal and marine systems. But, it also requires public awareness and participation, leading to investments in conservation, and the development of incentives, institutions and governance structures that will encourage individual resource users to respect the limits of the system within which they operate.

Most decisions to use coastal and marine resources are taken on the basis of private interests. If it is in the private interest of individuals to overexploit some system, the quality of scientific knowledge at their disposal will be irrelevant. Moreover, even the best scientific knowledge will leave considerable uncertainty about the future evolution of an ecological system.

● **To address the complexity of management regimes, it is essential to develop a methodology and collect the information required for the systematic valuation of ocean assets and services.**

One practical use of the attempts – as summarized in Table 1 – to estimate the value of ecological services, including the oceans, would be to modify systems of national income accounting to better reflect the contribution of these services to national and, thus, global income. Statistical offices of all countries should take up systematically the effort to estimate the value of such services, while correcting indices of GNP to take into account the pricing of ecological services and the depletion of natural assets.

The sustainable development of marine resources ultimately depends on establishing incentives that protect the resilience of exploited ecosystems and adapt to the flow of new information about the properties of the system. Given this, a combination of market signals and protective or precautionary measures is required at a scale that matches the

> The sustainable development of marine resources ultimately depends on establishing incentives that protect the resilience of exploited ecosystems and adapt to the flow of new information about the properties of the system.

115

structure and dynamics of the system being managed and is sensitive to social objectives.

The explosive increase in human activity, growing global interdependence, and rapid technological progress, are exerting a profound influence on the ability of ocean ecosystems to generate value for humankind. We need to address the issue of how to live with the oceans, which are today so much part of our home on planet earth. The economic challenge is how to exploit the oceans efficiently and equitably without compromising the interests of future users and consumers. Ocean governance requires management regimes that are robust and flexible enough to cope with sudden changes provoked by human actions, but also to contain the risk of sudden changes in the ecology of exploited systems. In the final analysis, the masters of the oceans are people, not the vagaries of the market place.

If people are to exercise their responsibilities for the prudent management and use of the oceans, they must possess the requisite knowledge as well as opportunities for influencing decision-making on the oceans. It is to these related issues of information and participation that we will now turn.

In the final analysis, the masters of the oceans are people, not the vagaries of the market place.

116

Environmental issues are best handled
with the participation of all concerned citizens...

Rio Declaration on Environment
and Development, Principle 10, 1992

Most of humanity is disenfranchised in the current
international system.

Philip Allott, *Mare Nostrum: A New International
Law of the Sea*, 1993

OUR OCEANS: PUBLIC AWARENESS AND PARTICIPATION

5

T he world's oceans are not only the domain of nation states and international organizations. They are also the legitimate concern of every human being. All those who have an interest or stake in the future of the oceans are natural and necessary participants in processes of ocean policy formulation and implementation.

Agenda 21 highlights the participatory role of major social groups in decision-making on sustainable development (section 3), the importance of public awareness and the availability of information to the public for this purpose (programme areas 36/B and 40/B) and, in the context of the integrated management and sustainable development of marine and coastal areas, the need to 'Provide access, as far as possible, for concerned individuals, groups and organizations to relevant information and opportunities for consultation and participation in planning and decision-making at appropriate levels' (objective 17.5).

Although there are welcome signs of change, opportunities for genuine civic–democratic involvement in ocean affairs remain very limited. This situation can only be corrected by major advances in public awareness and public participation.

PUBLIC AWARENESS, INFORMATION AND KNOWLEDGE

Public awareness of the oceans – the values they represent and the risks they face – is both a prerequisite for, and a result of, public participation in ocean governance. Beyond the simple human thirst for knowledge and quest for experience, people should know more about the oceans because they affect all of us profoundly, directly and indirectly, and in countless

119

ways. We must know more so we can use the oceans with respect and intelligent care, and safeguard them for future generations.

Improved information on the marine environment is dependent on the ability to convey a convincing message and, ultimately, to motivate decision-makers and the general public. The range of media available is vast, and varies in degree of sophistication and cost. It includes: the printed media (newspapers, literary works, information bulletins, newsletters, scientific journals, popular magazines, sign boards); the broadcast media (radio, television, telecommunications); lectures, public meetings, parades, demonstrations, fairs and expositions; films, audio cassettes and video clips, CDs and CD-ROMs; product labels, T-shirts, pins and badges; flags, banners, even hot air balloons and blimps. The Internet – a spectrum all its own – complements several features of 'conventional' media, and is of considerable importance because of its global volumes of information at negligible cost.

To be effective, any campaign must be sensitive to the audience targeted. In light of the message to be conveyed, certain media are to be preferred over others, depending on such characteristics as gender, age group, language, geographical area, occupation, social position,

Agenda 21 on participation

One of the fundamental prerequisites for the achievement of sustainable development is broad public participation in decision-making. Furthermore, in the more specific context of environment and development, the need for new forms of participation has emerged. This includes the need of individuals, groups and organizations to participate in environmental impact assessment procedures and to know about and participate in decisions, particularly those which potentially affect the communities in which they live and work. Individuals, groups and organizations should have access to information relevant to environment and development held by national authorities, including information on products and activities that have or are likely to have a significant impact on the environment, and information on environmental protection measures.

Source: Agenda 21, paragraph 23.2 (1992).

Importance of the sea for Brazilians
(from urban areas)

Brazilians went to the beach at least once: 77%

They consider the sea somehow important: 80%
- considerably important   66%
- of some importance   14%

The sea is important as:
- food resource   32%
- leisure   17%
- other natural resources   10%
- climate control   9%
- petroleum resource   7%
- transport   5%

Main worries concerning the sea:
- pollution of the beaches   56%
- uncontrolled fisheries activities   51%
- sea pollution   45%
- extinction of marine animals   44%
- nuclear experiments   36%
- rise of the sea-level   30%
- exploitation of the coast by foreigners   23%
- extinction of mangroves   9%

Brazilians think the majority of the beaches is polluted: 62%

Causes of beach pollution:
- litter thrown by users   45%
- urban sewage   29%
- oil residue   9%
- polluted rivers   2%

Pollution of the beaches is harmful to people: 91%
- only for the users   23%
- even for the non-users   68%

Mangroves are important: 58%

Endangered marine animals should be protected: 75%

Source: Instituto Gallup de Opinião Pública, for INCO-Brazil, August 1997.

ethnic identification, user/consumer interests in the oceans and ocean products, and financial capacity. If it is to work, the dialogue must be genuinely interactive, with the audience itself serving as a source of information, and it must be conceived in terms of a process based on shared interests and mutual benefits.

What constitutes the message is not simply built from words and ideas but also from personal experiences and images. A single incident, such as the *Torrey Cannon* oil spill or the sinking of the *Rainbow Warrior*, can trigger awareness leading to conviction and action that a decade of technical analysis of marine pollution could never accomplish. Dramatizing the importance of the oceans cannot ignore the fact that, to a great extent, the media are much better at telling real-life stories than at delivering abstract messages. Symbolism is also important in making the oceans an issue. The oceans are fresh and blue, not terrestrial green. Images such as a whale, dolphin or seal, an oil tanker, or satellite pictures of the oceans are all evocative symbols that are able to excite the imagination and can be powerfully joined to an awareness about life-supporting marine environments.

People can be reached not only collectively but also as individuals capable of furthering the cause of the oceans through their own actions, driven by ethical values and a sense of moral responsibility. The opportunity to act as an individual presents itself in the voting booth, in the course of one's daily routine, and in decisions taken in the market place. By setting an example to others, the acts of single individuals can have a magnifying influence and exert a powerful impact on collective behaviour. One illustration of a media-induced attempt to tap the individual consumer's sense of responsibility is the Marine Stewardship Council sponsored jointly by WWF and Unilever (see Chapter 2).

Our children urgently need to be educated about the oceans. Little could be more important than passing on to future generations a knowledge of the myriad creatures that live in the oceans and of the relationships between the seas and life on land. Changing the attitudes of the young towards the oceans can build public support for improving the marine environment. The main objectives of educational programmes should be to sensitize children and young people to the oceans – to the role they have played in the history of humankind, in the birth and formation of civilization, and in processes of social change and economic progress. They must also sensitize young people to environmental

> Symbolism is also important in making the oceans an issue. The oceans are fresh and blue, not terrestrial green.

UNESCO Associated Schools –
Caribbean Sea Project

The UNESCO Associated Schools – Caribbean Sea Project (CSP) is a 'flagship project' designed to promote a culture of peace. Among other UNESCO Associated Schools regional activities are the Baltic Sea Project, the Danube River Project and the South Eastern Mediterranean Sea Project.

Participants: The CSP was launched in Trinidad and Tobago in November 1994. Seven countries and territories participated in the first phase of the project: Cuba; Curaçao (all of the Netherlands Antilles from 1996); Grenada; Jamaica; Saint Vincent and the Grenadines; Trinidad and Tobago; and Venezuela. Nine more countries and territories joined in the second phase (1996–1997): Aruba; Bahamas; Barbados; Cayman Islands; Colombia; Costa Rica; Dominica; Haiti; and Saint Lucia. By the end of 1998, it is hoped that all countries and territories around the Caribbean Sea will be participating. The Secretary General of the Trinidad and Tobago National Commission for UNESCO serves as Regional Coordinator for the project.

Goals: The overall goal of the CSP is to protect and preserve the Caribbean Sea and its waterways as a regional patrimony for sustainable human development and to strengthen Caribbean identity. The specific objectives of the project are:

(i) to sensitize children and young people to environmental problems facing the Caribbean Sea both locally and regionally, and to develop their skills in helping to resolve these problems;

(ii) to develop new interdisciplinary/multidisciplinary educational approaches and materials to involve children and young people in dealing with problems threatening the marine environment;

(iii) to provide a mechanism for interaction, exchanges and cooperation amongst schools from different language groups.

Source: UNESCO (1997).

problems facing the oceans, and contribute to the development of the knowledge and skills required to resolve them.

Because of the respect they enjoy in most societies, scientists, particularly marine scientists, can contribute significantly to an educational process by preparing audio-visual materials and experiments for school teachers, monitoring school trips to an aquarium or to the sea, and providing role models for children and young people in relation to the marine environment. Support from the business and academic communities should be solicited to organize outreach programmes and to develop educational materials, making use of multimedia techniques and the Internet, that go beyond the conventional classroom in providing young minds with the knowledge needed to protect marine resources as well as preparing them for exciting ocean-related careers.

The Biodiversity Convention and Agenda 21 recognize the role of indigenous and local communities in the protection and sustainable use of marine resources and biological diversity. Pacific islanders, for example, are known to have devised and practised almost every basic form of marine conservation measure centuries before the need for marine conservation was even recognized by the industrialized countries. It is necessary to identify, protect and diffuse this knowledge, including innovations and practices which have proven their effectiveness in the regulation and control of marine habitats in a variety of situations. The media can be effective in disseminating this ancient ecological wisdom, as well as information about the benefits of traditional technologies.

Public perceptions are closely linked to relevant experience. In general, people do not perceive environmental problems in an abstract manner, but in a local or contextual setting. More often than not, public awareness is linked to the perception of interests, or of risks, associated with the use of the immediate environment and its resources.

The growth of scientific knowledge and technological development, combined with increasing appropriation of natural resources, has contributed to expanding global awareness, especially in affected sectors of society, about the role of the ocean system both as a provider of new living and non-living resources and as a crucial factor in the earth's climate. A greater perception of the risks, stemming from uncontrolled uses of the oceans, has also developed. However, this consciousness tends to be confined to those who – like marine scientists,

Pacific islanders' perceptions of the ocean

For Pacific societies for whom writing was unknown until the post-contact period, navigation information upon which the life of crew and navigator depended was passed on through chants which were easily memorised because of their rhythmic and verse-like qualities. Much of the movement across the wide expanse of ocean by people of Oceania was preserved and shared in this way.

The eastern horizon

The handle of my steering paddle thrills to action,

My paddle named Kautu-ki-te-rangi.

It guides to the horizon but dimly discerned

To the horizon that lifts before us,

To the horizon that ever recedes,

To the horizon that ever draws near,

To the horizon that causes doubt,

To the horizon that instils the dread,

To the horizon with unknown power,

The horizon not hitherto pierced.

The lowering skies above,

The raging seas below,

Oppose the untraced path

Our ship must go.

Polynesian Deep-sea Chanty.

fishery and coastal communities, or off-shore oil developers – have a direct or indirect interest in the use and/or in the protection and preservation of the oceans.

Information – the basis for knowing

A major step in expanding public awareness about the oceans, their wealth and fragility, is to ensure wide availability and dissemination of relevant knowledge and information. Information and knowledge currently depend on scientific expertise. Information conveyed by the mass media is based, most often, on scientific or technical sources. However, informed public participation in decision-making on the oceans requires better public understanding of science, including the extent and the limitations of existing knowledge about the oceans. Moreover, there is an increasing need for broadening the scope of information so as to cover the relevant socio-economic aspects underlying policy-making and management. A special effort should be made to present analyses and assessments to the public in an accessible form, and to increase the transparency of consultative and decision-making procedures.

The belief in science as the authoritative source of knowledge, which underpins Western culture, has suffered some setbacks in recent times, as disagreements among scientists, and uncertainties which characterize the development of science, are increasingly recognized. Quite often, the perceptions of environmental phenomena by ordinary people, despite their unsystematic character, can contribute relevant information to processes of policy formulation, especially by drawing attention to ethical considerations falling outside the confines of science.

A recurrent theme in recent debates, exemplified by the Rio Earth Summit as well as by the Special Session of the UN General Assembly held in June 1997, is the need for appropriate action under conditions of uncertainty. Uncertainties arise because relevant data are lacking, the systems for acquiring data do not exist or function well, the technology for making the necessary observations has not yet been perfected, and experts disagree. Uncertainties are commonly perceived because the natural or social systems we are seeking to predict, manipulate or protect are exceedingly complex. Some such systems – not necessarily especially complex ones – are, it is increasingly accepted, inherently unpredictable and their results are subject to contradictory interpretations. The different kinds of uncertainties need to be approached in different ways, in the scientific, legal and socio- economic realms, and in light of the

...informed public participation in decision-making on the oceans requires better public understanding of science, including the extent and the limitations of existing knowledge about the oceans.

'precautionary approach' endorsed in Principle 15 of the Rio Declaration (as discussed in Chapter 4).

While sectors of our societies struggle with this matter, it is important that those concerned with raising awareness and providing information to the public neither oversimplify the issues nor render them in confusing and needlessly complex ways. Scientific findings and associated pronouncements identify uncertainties, and these often lead to controversy within the scientific community. There are even some dangers of misunderstandings when scientists attach probabilities to their conclusions, as they do more and more frequently; numerical probabilities themselves depend on assumptions in the methods used to estimate them. However, this should not lead to the dangerous conclusion that science has little value in such circumstances. The existence of inherent, unresolvable uncertainties in the behaviour of systems, such as the ocean–atmosphere climate system, does not mean that we cannot hope to cope with the vagaries of those systems better with than without science. The terminology of 'fractals', 'chaos', 'complexity', 'self-similarity' needs to be explained so that ordinary people can grasp the underlying ideas. The language of the simulation of natural and man-made systems with computers, which is becoming so important in the management of marine activities, also needs to be explained in a popular form. These concerns identify important and fruitful fields for interaction, mutual assistance and support among natural scientists, mathematicians, logicians and linguists, economists, social scientists, lawyers, and educators.

The requirements for messages about the oceans – their nature and uses, and the threats to them – are enormous and diverse. It will be essential to create a universal information base, and to agree on the institutional and other arrangements for maintaining and improving it, and on the rules conducive to democratic access. It is important to understand the vital role played by such an information base in this 'information age'. For most of the second half of the twentieth century, oceanographers and others have been committed to the creation of national, regional and global data centres. Most of the centres established were specialized in scope. They were progressively connected with each other and gradually assumed the form of a globally extensive but incomplete network. The logistical problems remain enormous, and often seem intractable.

The explosive spread of digital processing in practically all fields of human activity, and the

Digital information networks

Digital information networks have four main characteristics.

First, and most obviously, they hold, move and generally handle information in digital form.

The second characteristic is that they require only minimal management; in the Internet, for example, it is only necessary to establish an agreed system for naming sites, nodes and locations, and to arrange for the custody and improvement of the language used by millions of computers in their interactions.

The third characteristic is that digital information can be translated, via other media, into multimedia products: written and spoken words, still and moving coloured graphic images, music and other sounds. Transmission of our sense of touch is not far away, and perhaps also even our senses of taste and smell.

The fourth characteristic of digital networks – which is perhaps the most important here – is the ease with which internal connections can be made. The best known and most widely used implementation of this ability is the World Wide Web. Apart from the multimedia content of sites on the Web, the abilities to link sites, and to develop search engines capable of wandering like robots throughout the Web, offer us the possibility of a universal, virtually unlimited information system that operates continuously without a centre.

Numerous Web sites concerned primarily with the oceans are already in existence. There are also numerous sites with wider scope but containing marine elements. Examples among international NGOs include the sites maintained by Greenpeace International, WWF and IUCN. National and international institutions also maintain a burgeoning number of Web sites. No count has yet been made, but such sites of more or less direct interest to ocean affairs already number in the hundreds and are increasing daily.

Source: Sidney Holt (1997).

construction and growth of the Internet, have changed the situation dramatically. We should no longer think of knowledge centres, but rather of open networks. Such networks possess many of the features of living organisms. They exhibit self-organization; and they evolve in unpredictable ways. Whether or not the Internet itself continues to exist in its present form, global electronic 'digital information networks' are here to stay and our ability to develop world consciousness of ocean affairs depends largely on our ability to use them effectively. That ability, too, is unevenly distributed in the world today, and hence should be addressed by special efforts at bridging the gaps, as previously mentioned in Chapter 2.

If wider public participation in ocean affairs is to become a reality, a comprehensive strategy for public awareness should be developed to support it. Current international communications technologies offer an unprecedented opportunity to raise public awareness about the oceans as part of our global heritage and common concern. Being the International Year of the Ocean, 1998 constitutes the historic moment to seize this opportunity.

PUBLIC PARTICIPATION

The recent emergence of new theories and practices of public participation is closely related to the recognition of a general interest by non-state social groups in the environment and natural resources. This interest is combined with growing scepticism regarding the ability of governments to resolve environmental problems solely on the basis of technical expertise, especially given the pressures exerted on governmental policy by private enterprise.

Policy-making in the economic and social spheres has increasingly become an affair of the executive branch of governments, rather than of parliaments, which have lost a significant part of their former power as political and administrative decision-making became more and more technical and bureaucratized. This phenomenon is leading to a re-thinking of the foundations and mechanisms of democracy. As a consequence, new forms of legitimation of administrative authority are emerging, including the advocacy of greater transparency and more openness to external participation in the decision-making process. The involvement of professional, economic and social stakeholders in governance is a growing trend in democracies.

Ocean governance, and the management of territorial waters and EEZs in particular, is in this sense

The right to
participate
in ocean
governance
thus implies the
acceptance of,
and effective
compliance
with, the duty
to respect the
rights of others,
neighbours,
humanity as a
whole, and
future
generations.

paradigmatic: the public, non-proprietary character of these marine areas makes them especially amenable to participatory forms of management. As observed in the 1995 Report of the Commission on Global Governance, national civil societies have begun to merge into a wider global civil society. Rights should be thought of in broader terms than those of the relationship between people and governments. Additionally, rights must be combined with duties. The right to participate in ocean governance thus implies the acceptance of, and effective compliance with, the duty to respect the rights of others, neighbours, humanity as a whole, and future generations. These considerations should drive the formation of inter-national legal provisions and mechanisms designed to empower people to participate in ocean affairs.

The 1987 report of the World Commission on Environment and Development has already set forth three ways of ensuring participation by non-state entities in environmental decision-making: 'their right to know and to have access to information on the environment and natural resources; their right to be consulted and to participate in decision-making on activities likely to have a significant effect on their environment; and their right to legal remedies and redress when their health or environment has been or may be seriously affected'.

While many mechanisms exist to make participatory rights operational at the national level, particularly in the environmental field, they are still poorly reflected in current international ocean agreements and practice.

The right to know – and the means to know

Few international provisions require public disclosure of information regarding activities in ocean space, within or beyond national jurisdiction. Although several regional or global marine agreements refer to the communication, publication and dissemination of information, treaties have traditionally been more concerned with scientific and interstate exchanges than with the public's right to know. Only a few recent agreements even address the need for public disclosure of relevant information: the 1991 Madrid Protocol on Environmental Protection to the Antarctic Treaty (articles 14 and 17), for example, expressly provides for public availability of inspection reports and annual reports.

The 1992 Convention for the Protection of the Marine Environment of the North-East Atlantic (OSPAR, article 9) calls on the Contracting Parties to ensure that their competent authorities are required to make information available 'to any natural or legal person, in response to any reasonable request, without that person having to prove an interest, without unreasonable charges, as soon as possible and at the latest within two months'.

The 1995 amendments to the Barcelona Convention for the Protection of the Marine Environment and the Coastal Region of the Mediterranean (article 11B) require Contracting Parties to ensure that their competent authorities shall give to the public 'appropriate access to information on the environmental state in the field of application of the Convention and the Protocols, on activities or measures adversely affecting or likely to affect it and on activities carried out or measures taken in accordance with the Convention and the Protocols'.

The 1995 Agreement on Straddling Fish Stocks and Highly Migratory Fish Stocks (article 12) calls for 'transparency in the decision-making process and other activities of sub-regional and regional fisheries management organizations and arrangements', and provides that intergovernmental and non-governmental organizations 'shall have timely access to the records and reports of such organizations and arrangements, subject to the procedural rules on access to them'.

At the same time, the activities of international organizations are beginning to be subjected to similar requirements. For example, the World Bank's 1993 Procedure for Disclosure of Operational Information (BP 17.50), which triggered parallel enactments by regional financial institutions (such as the 1994 Information Disclosure Policy of the Inter-American Development Bank), also applies to projects implemented by the World Bank under the Global Environment Facility (GEF). UNDP and UNEP in turn issued administrative 'policies and procedures related to public availability of documentary information on GEF operations' in September 1993.

In practice, however, international disclosure rules are subject to severe limitations aimed at safeguarding confidentiality, security and other interests. Echoing the 1990 EC Council's Directive on Freedom of Access to Information on the Environment (90/313/EEC), the 1992 OSPAR Convention thus reserves the right of governments to refuse disclosure 'where it affects: the confidentiality of the proceedings of public authorities, international relations and national defence; public security; matters which are, or have been, sub judice, or under enquiry (including disciplinary enquiries), or which are the subject of preliminary investigation proceedings; commercial and industrial confidentiality, including intellectual property; the confidentiality of personal data and/or files; material supplied by a third party without that party being under a legal obligation to do so; material, the disclosure of which would make it more likely that the environment to which such material related would be damaged'.

Another potential constraint on public access to information is the frequent delegation of resource management functions to parastatal or non- governmental entities, especially in the context of privatization. Consequently, the Council of Europe's 1993 Convention on Civil Liability for Damage Resulting from Activities Dangerous to the Environment (articles 15 and 16) provides access to information held not only by public authorities but also by 'bodies with public responsibilities for the environment and under the control of a public authority', and to specific information held by operators.

Equally significant, especially for developing country participants, are financial constraints on information access. Among the means to overcome these obstacles, and to facilitate broadly inclusive information-sharing, are provisions to ensure periodic publication and dissemination of reports, user-friendly documentation services and – where feasible – open availability through the Internet, such as the World Bank's Public Information Center established in 1994. It must be realized, however (taking into account the limited resources of the international agencies concerned), that transparency and accountability also have their price.

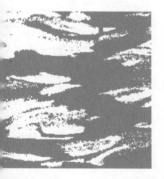

Last, but not least, among traditional shackles on the right to know is its limitation to persons with 'legal capacity' to exercise it – hence excluding not only those still unborn (i.e. future generations) but also those still below the recognized legal age (i.e. minors). Yet, if there is anybody who ought to be fully informed about the state of the oceans they will inherit, it is our children – and those who, even after them, will have to live with the environmental time-bombs we are leaving behind. If the principle of intergenerational equity (re-affirmed in the 1997 UNESCO Declaration on the Responsibilities of Present Generations Towards Future Generations) is to be given effect, new forms of fiduciary representation will have to be found for these future constituencies.

The right to be heard – and the means to be heard

Public participation – understood as a civic right, independent of a specific stakeholder interest – may be justified on moral grounds (the democratic ideal implying that people share in governance), on grounds of efficiency (the contribution of social learning to improved decision-making), and on the build-up of trust in public institutions that makes their decisions socially acceptable and politically legitimate. The majority of participatory procedures in ocean affairs are scientifically motivated by the rationale of effectiveness; only in a few cases is the participation of non-state entities expressly based on reasons of wider democratic representation and consultation.

Historically, the issue is linked to the admission of non-state observers in the international decision-making bodies concerned. The gradual expansion of NGO representation at meetings of the International Whaling Commission, the IMO Marine Environment Protection Committee and the London Dumping Convention, in particular, is well documented. Other examples confirming this trend are the growing role of the ICSU Scientific Committee on Antarctic Research (SCAR) and the Antarctic and Southern Ocean Coalition (ASOC) of NGOs in the context of the Antarctic Treaty regime, and NGO attendance, from 1990 onwards, at the regional commissions for marine environment protection in the Baltic Sea (HELCOM) and the North-East Atlantic (OSPARCOM), and the related ministerial conferences.

The 1992 Rio Earth Summit, attended by more than 1400 NGOs, triggered a review and gradual liberalization of the existing rules on NGO observer status in the UN system, as originally laid down by the UN Economic and Social Council (pursuant to article 71 of the UN Charter) in 1968 and now revised by ECOSOC Resolution 1996/31 of 25 July 1996. It also opened the door for procedural changes in the Bretton Woods institutions, with the Council of the GEF permitting, since 1995, a limited number of NGO observers to attend sessions.

While the Law of the Sea Convention (article 169) had merely called on the International Seabed Authority to 'make suitable arrangements for consultation and cooperation' with NGOs recognized by ECOSOC, the 1995 Fish Stocks Agreement (article 12/2) provides that representatives from NGOs concerned with straddling fish stocks and highly migratory fish stocks

Public participation... may be justified on moral grounds ... on grounds of efficiency... and on the build-up of trust in public institutions that makes their decisions socially acceptable and politically legitimate.

Public
participation
should
encompass
not only the
process during
which decisions
are made, but
also the
subsequent
process of
implementation
– including
complaints
and remedies
against the
wrong type
of decisions.

'shall be afforded the opportunity to take part in meetings of sub-regional and regional fisheries management organizations and arrangements as observers or otherwise, as appropriate, in accordance with the procedures of the organization or arrangement concerned. Such procedures shall not be unduly restrictive in this respect'. At the regional level, the 1995 amendments to the Mediterranean Convention (article 11B/2) require the Contracting Parties to 'ensure that the opportunity is given to the public to participate in decision-making processes relevant to the field of application of the Convention and the Protocols, as appropriate'. Despite these gains, significant limitations remain. Under the 1992 OSPAR Convention (article 11), any member state may veto the participation of an NGO. This rule is more restrictive than provisions in numerous other global environmental treaties, which require objection by at least one-third of the governmental parties represented to exclude an NGO.

Moreover, procedural rules under a number of instruments – for example, article 14 of the 1994 GEF Rules of Procedure – reserve the right to exclude the public, and even accredited observers, from 'closed' (i.e. intergovernmental) sessions. Observer status thus does not necessarily confer a right to be heard, let alone a right to affect decisions, which typically depends on internal procedures and/or to a large extent on the chairperson's discretionary powers.

By the same token, an observer's capacity to participate may be limited by financial factors, such as travel and accreditation costs. In recognition of these constraints, and especially of the North–South imbalance which they tend to exacerbate, several international institutions (for example, GEF) have made a special provision to support the attendance of NGO observers from developing countries, through voluntary trust fund contributions.

The right to complain – and the means of recourse

Public participation should encompass not only the process during which decisions are made, but also the subsequent process of implementation – including complaints and remedies against the wrong type of decisions. At the national level, principle 10 of the Rio Declaration calls for 'effective access to judicial and administrative proceedings, including redress and remedy', and in the specific case of marine pollution damage, the Law of the Sea Convention (article 235/2) requires states to 'ensure that recourse is available in accordance with their legal systems for prompt and adequate compensation or other relief'.

Internationally, however, most legal remedies are still restricted to governmental claimants: neither the International Court of Justice (ICJ Statute, article 34) nor the dispute settlement understanding of the World Trade Organization (WTO Agreement, annex 2) are open to entities other than states; and the International Tribunal for the Law of the Sea will accommodate non-governmental entities as parties to proceedings only in connection with seabed exploration or exploitation (the Convention, articles, 153, 187 and 190).

International arbitration mechanisms are not so limited – as illustrated by the Permanent Court of Arbitration's 'Optional Rules for Arbitrating Disputes Between Two Parties of Which Only One Is a State' (The Hague, 1993). However, submission to arbitration always depends on consent by the parties concerned, including a waiver of a state's sovereign immunity. Only under rather exceptional circumstances – such as the Rainbow Warrior arbitration (Greenpeace vs. France, Geneva 1987, where the 'delinquent' government waived its immunity) – would non-governmental entities gain access through that option.

Another standing limitation is the traditional requirement for claimants to show that they are directly concerned or affected by the decision against which recourse is sought. Even under the rules of the European Court of Justice, which permit private actions against Commission decisions (EC Treaty, article 173), a claim by Greenpeace was held inadmissible for lack of individual concern in relation to the environmental consequences of a contested decision (Canary Islands case, 1996). The same jurisdictional limitation applies in the otherwise innovative procedure of the World Bank's Inspection Panel established in 1993: an affected party (such as an organization, association, society or other grouping of individuals) must demonstrate that its rights or interests 'have been, or are likely to be, directly affected' by the contested decision or project. In the case of alleged harm to the oceans, it is notoriously difficult to provide such proof – especially where it concerns areas beyond national jurisdiction.

There appears to be only one instance of an international complaint mechanism that would allow public recourse against environmentally harmful state action regardless of direct harm to a claimant. In the European Union (EU), a complaints registry has been established whereby any citizen or civic group may complain about alleged infringements of EU environmental directives

On the whole, despite some promising initiatives, the reality of public participation in ocean governance today remains a far cry from the ambitious expectations created by Agenda 21.

by member states, without requiring the complainant to show direct harm. The EU Commission may then – as custodian of treaty implementation (article 155) – initiate proceedings against the member state, sanctioned, if necessary, by formal action in the European Court of Justice (article 169). If the Commission fails to act, the complaint may still be taken to the European Parliament's 'ombudsman'. It must be kept in mind, however, that this international mechanism is limited to the circumstances of the quasi-federal structure of the EU and the unique custodial role assigned to the Commission by its constitutive instrument.

On the whole, despite some promising initiatives, the reality of public participation in ocean governance today remains a far cry from the ambitious expectations created by Agenda 21. One reason for the disappointing record is the lack of independent representative institutions qualified to speak out on behalf of non-utilitarian transnational values – on behalf of civil society, future generations, the global environment, or of other public interests.

Proposals to bridge this gap are formulated in the chapter which follows. Meanwhile, however, continuous efforts are needed to strengthen the public's right to participate in ocean affairs – nationally and inter-nationally – to ensure that all states are held accountable for their management and conservation of the ocean resources entrusted to them for the benefit of all people. At the national level, many countries have created an office of 'citizen's defender' or 'guardian/custodian of citizen's rights', modelled after the 'ombudsman' originally established in Sweden under its 1809 Constitution and later, by other Scandinavian countries to safeguard citizens against governmental abuses. Similar offices now exist in more than 80 states throughout the world, sometimes in the form of a 'parliamentary commis-sioner' (New Zealand) or a 'permanent commission of enquiry' (Tanzania), with wide-ranging powers of inspection and access to government-held information, either in response to citizens' complaints (reactive) or independently initiated (pro-active). A regional ombudsman's office has been established in the European Parliament (article 138e pursuant to the 1992 Maastricht Treaty) to deal with grievances against inter-governmental European institutions.

At the global level, consideration should be given to the appointment of an independent Ocean Guardian, with a mandate to take up grievances concerning alleged non-compliance with international marine agreements or misuse of the oceans and their resources.

Keeping in mind the need to safeguard the incumbent's independence and impartiality (which at the national level is normally guaranteed by parliamentary designation and accountability of an ombudsman), appointment by an appropriate intergovernmental institution with global authority in ocean affairs (such as the UN General Assembly) would be desirable. An Ocean Guardian's office should in no way duplicate, but effectively supplement, existing means of formal recourse at the national or international level. It would have to be equipped to receive and investigate complaints, and to report its findings to the competent institutions with a view to resolving issues in a non-adversial manner. Its principal function would thus be to defuse conflicts, and to facilitate the interaction of non-state entities with the intergovernmental system.

A more informed and active civil society with significantly enlarged opportunities to participate in ocean affairs is a precondition for a more responsive and democratic system of ocean governance. It is in the critically important area of ocean governance that competing issues and divergent interests and opinions must be accommodated and reconciled. We will now turn to ocean governance and, in doing so, we will seek to weave together the threads that run through this report.

'There is nothing more difficult to carry out, nor more doubtful
of success, nor more dangerous to handle,
than to initiate a new order of things.'

Machiavelli, *The Prince*, 1513

'Effective global governance calls for a new vision, challenging
people as well as governments to realize that there is no
alternative to working together to create the kind of world
they want for themselves and their children.'

Commission on Global Governance, *Our Global
Neighborhood*, 1995

TOWARDS EFFECTIVE OCEAN GOVERNANCE

6

In this report, we have looked at many of the challenges that confront us in the oceans. While acknowledging the progress that has been recorded in a number of areas, and the importance of numerous new initiatives linked to the sustainable use and management of the oceans and its resources at different levels, we have been compelled to conclude that the global community still lacks an effective system of governance capable of ensuring that the oceans are used for the benefit of all and in the interests of future generations.

It is true that each of the past three decades made its own contribution to the emergence of an improved understanding of the relationships existing between the oceans and the long-term conservation of our natural heritage. Beginning with the Stockholm Conference on the Human Environment in 1972, growing concerns over pollution and depletion of marine resources gained voice, leading to the establishment of yet another variety of organizational structures. Environmental concerns were acknowledged in the UN General Assembly's call for a new regime of ocean law, with responsible stewardship and the equitable utilization of ocean resources as major goals.

The next decade witnessed the adoption of the UN Convention on the Law of the Sea in 1982 aimed at the establishment of a new legal order for the oceans, the promotion of the equitable and efficient utilization of ocean resources, and the preservation of the marine environment.

Ten years later in 1992, the UN Conference on Environment and Development (UNCED) translated many of the concerns expressed into a more coherent vision of sustainable development aimed at promoting the social and economic well-being of present and future generations in ways consistent with the conservation of the earth's resources and natural systems. The principle of 'common but differentiated responsibilities' was endorsed to advance equity as well as cooperative international initiatives in the marine realm (Chapter 17 of Agenda 21).

Over the past three decades, the international community has thus been involved in permanent negotiations that seek to establish an improved legal framework for the use of the oceans and for the more equitable management and conservation of marine resources. Throughout this process of negotiation, there has been recognition of the need to create an appropriate institutional framework, but this recognition has so far failed to result in a coherent system of ocean governance.

The problem today is how to transform an aggregate of sectoral institutions existing at the national and international levels into a flexible and dynamic network that is responsive to the goals of solidarity and sustainability and to our growing knowledge of ecological linkages. The policies adopted by governments, the administrative institutions they establish, and the information and analysis so vital to sound management, remain fragmented among different ministries and agencies at the national level, and among different international institutions at regional and global levels. This fragmentation makes it difficult to reconcile competing uses and adds greatly to the difficulties of defining well-founded priorities for the oceans and coastal areas. It has the effect of truncating ocean governance into piecemeal and short-sighted measures, and it perpetuates a vicious cycle of overuse and environmental degradation. This further undermines the well-being of society and increases the potential for internal and international conflicts.

As the number of international legal instruments multiplies, it has become more difficult for governments and the general public to keep track of the steady flow of new commitments that stem from both legally binding conventions and non-binding codes, guidelines and strategies.

Given this situation, it is tempting to think in terms of a new 'blueprint' for the oceans that provides for the creation of powerful new institutions, backed by the force of international law, that are mandated and provided with the authority to manage the oceans in ways that promote peace, security and justice and safeguard the interests of the unborn. While a case could undoubtedly be made for such a blueprint, it is likely to have little relevance outside of academic debate. The oceans form an arena of conflicting uses and interests which often preclude the political consensus required for bold exercises in institutional innovation, desirable as they may be. Even so, the existing institutional framework offers a number of opportunities for action now.

The problem today is how to transform an aggregate of sectoral institutions existing at the national and international levels into a flexible and dynamic network that is responsive to the goals of solidarity and sustainability and to our growing knowledge of ecological linkages.

POSITIVE DEVELOPMENTS

It is essential that strategies for the evolution of ocean governance take account of positive developments, which can be defined to include the following:

● The oceans are increasingly recognized as a vital component in the hydrological, geophysical, and atmospheric systems that impact on levels of well-being and welfare and the habitability of our planet.

● The increasing involvement of civil society as a whole in ocean affairs at all levels has given impetus to the process of change. Governments have begun to call on civil society in attempts to reconcile conflicts in the oceans. This is encouraging greater awareness within society of the role of stakeholders in ocean policy and of the role of both national and international arrangements in articulating their interests in achieving policy objectives.

● There are legal foundations for the promotion of greater equity in the oceans, notably the concept of the 'common heritage of mankind'. Now applied to the deep seabed and its mineral resources, the concept establishes a model for use in other areas, such as the genetic resources of the deep seabed.

● The combination of terrestrial and marine uses in the coastal zone has focused new attention on integrated coastal zone management, and has compelled national and international bodies to assess their disparate approaches and guidelines.

● There is growing recognition of the need to think inclusively in terms of systems that encompass both sea water and fresh water. Recognition of the inter-dependencies existing between fresh water drainage basins, wetlands and the coastal zone is giving rise to new international initiatives in integrated watershed management which take into account impacts on coastal and marine areas (such as the Black Sea and Danube basin programmes).

● The ecosystem-based and precautionary approaches to the management of living resources have spread from the regional level, for example, for Antarctic fisheries, to the global level, reflected, for example, in the 1995 Agreement on Straddling Fish Stocks and Highly Migratory Fish Stocks.

● At the policy level, the linkages between international legal arrangements for the oceans, biodiversity, wetlands, rivers, the atmosphere, movements of hazardous wastes, and other related issues have been explicitly incorporated into some regional arrangements, notably for the Mediterranean.

● At the operational level, the restructured Global Environment Facility (GEF), a semi-autonomous funding mechanism that draws on the resources of multilateral and bilateral bodies, has begun to develop an integrated International Waters programme for marine and fresh water resources.

● In the field of information, efforts are being made to identify and organize the data required by governments to diagnose ocean problems and to find solutions to them. Of particular importance is the Global Programme of Action for the Protection of the Marine Environment from Land-Based Activities, adopted in Washington, in 1995, which has established a clearing house mechanism under UNEP auspices, with the prospect of linkages to the marine and coastal bio-diversity issues addressed by the Convention on Biological Diversity.

● The growing emphasis placed on the value of natural resources and environmental services, and on the role of economic incentives, has stimulated the formulation of policies and programmes that promote the more rational use of ocean resources.

These positive developments point to a growing momentum for change. Although it remains to be seen whether they are together sufficient to overcome the obstacles rooted in vested interests and traditional sectoral approaches, they nevertheless set the stage for the adjustments required.

The oceans in Agenda 21

Agenda 21: Programme of Action for Sustainable Development was adopted, together with the Rio Declaration on Environment and Development, at the UN Conference on Environment and Development held in Rio de Janeiro, on 3–14 June 1992. Its Chapter 17 deals with the Protection of the Oceans, All Kinds of Seas, Including Enclosed and Semi-Enclosed Seas, and Coastal Areas and the Protection, Rational Use and Development of Their Living Resources. It states:

The marine environment – including the oceans and all seas and adjacent coastal areas – forms an integrated whole that is an essential component of the global life-support system and a positive asset that presents opportunities for sustainable development. International law, as reflected in the provisions of the UN Convention on the Law of the Sea... sets forth rights and obligations of States and provides the international basis upon which to pursue the protection and sustainable development of the marine and coastal environment and its resources. This requires new approaches to marine and coastal area management and development, at the national, sub-regional, regional and global levels, approaches that are integrated in content and are precautionary and anticipatory in ambit, as reflected in the following programme areas:

(i)   Integrated management and sustainable development of coastal areas, including exclusive economic zones;
(ii)   Marine environmental protection;
(iii) Sustainable use and conservation of marine living resources of the high seas;
(iv)   Sustainable use and conservation of marine living resources under national jurisdiction;
(v)   Addressing critical uncertainties for the management of the marine environment and climate change;
(vi)   Strengthening international, including regional, cooperation and coordination;
(vii) Sustainable development of small islands.

The implementation by developing countries of the activities concerned shall be commensurate with their individual technological and financial capacities and priorities in allocating resources for development needs and ultimately depends on the technology transfer and financial resources required and made available to them.

Source: Agenda 21: Programme of Action for Sustainable Development (1992).

United Nations Convention on the Law of the Sea

The Convention comprises 320 articles and nine annexes, governing all aspects of ocean space, such as delimitation, environmental control, marine scientific research, economic and commercial activities, transfer of technology and the settlement of disputes relating to ocean matters. The Convention entered into force on 16 November 1994, 12 months after the date of deposit of the sixtieth instrument of ratification or accession. Some of its key features are the following:

● Coastal States exercise sovereignty over their *territorial sea* which they have the right to establish up to a limit not to exceed 12 nautical miles; foreign vessels are allowed 'innocent passage' through those waters;

● Ships and aircraft of all countries are allowed 'transit passage' through straits used for international navigation; States bordering the straits can regulate navigational and other aspects of passage;

● *Archipelagic States*, made up of a group or groups of closely related islands and interconnecting waters, have sovereignty over a sea area enclosed by straight lines drawn between the outermost points of the islands; all other States enjoy the right of archipelagic passage through such designated sea lanes;

● Coastal States have sovereign rights in a 200-nautical mile *exclusive economic zone (EEZ)* with respect to natural resources and certain economic activities, and exercise jurisdiction over marine science research and environmental protection;

● All other States have freedom of navigation and overflight in the EEZ, as well as freedom to lay submarine cables and pipelines;

● *Land-locked and geographically disadvantaged States* have the right to participate on an equitable basis in exploition of an appropriate part of the surplus of the living resources of the EEZs of coastal States of the same region or sub-region; highly migratory species of fish and marine mammals are accorded special protection;

● Coastal States have sovereign rights over the *continental shelf* (the national area of the seabed) for exploring and exploiting it; the shelf can extend at least 200 nautical miles from the shore, and more under specified circumstances;

● Coastal States share with the international community part of the revenue derived from exploiting resources from any part of their shelf beyond 200 miles;

• The Commission on the Limits of the Continental Shelf shall make recommendations to States on the shelf's outer boundaries when it extends beyond 200 miles;

• All States enjoy the traditional *freedoms* of navigation, overflight, scientific research and fishing *on the high seas*; they are obliged to adopt, or cooperate with other States in adopting, measures to manage and conserve living resources;

• The limits of the territorial sea, the exclusive economic zone and continental shelf of *islands* are determined in accordance with rules applicable to land territory, but rocks which could not sustain human habitation or economic life of their own would have no economic zone or continental shelf;

• States bordering *enclosed or semi-enclosed seas* are expected to cooperate in managing living resources, environmental and research policies and activities;

• *Land-locked States* have the right of access to and from the sea and enjoy freedom of transit through the territory of transit States;

• States are bound to prevent and control *marine pollution* and are liable for damage caused by violation of their international obligations to combat such pollution;

• *All marine scientific research* in the EEZ and on the continental shelf is subject to the consent of the coastal State, which in most cases must be granted to other States when the research is to be conducted for peaceful purposes and fulfils specified criteria;

• States are bound to promote the *development and transfer of marine technology* 'on fair and reasonable terms and conditions', with proper regard for all legitimate interests;

• States Parties are obliged to settle by peaceful means their *disputes* concerning the interpretation or application of the Convention;

• Disputes can be submitted to the International Tribunal for the Law of the Sea established under the Convention, to the International Court of Justice, or to arbitration. Conciliation is also available and, in certain circumstances, submission to it would be compulsory. The Tribunal has *exclusive* jurisdiction over deep seabed mining disputes.

Source: UN Division for Ocean Affairs and the Law of the Sea (1997).

BARRIERS TO GOOD OCEAN GOVERNANCE

Multiple international institutions

Some of the customary legal rules governing the uses of the oceans have evolved over centuries through time-honoured processes of claim, counter-claim and eventual consensus among maritime and coastal nations. This evolving body of law has been able to draw, directly and indirectly, upon the traditions and experience not only of Western Europe but also of many other regions, such as Asia, East Africa and Polynesia. However, most contemporary ocean law, and especially its organizational framework, is a product of the second half of the twentieth century.

Most of the present body of rules and institutions has been developed since the 1950s in response to such challenges as resource depletion, scarcity and degradation, as well as in recognition of the value of scientific research, requiring increased international cooperation. The guiding objective in this process has been the need to avert conflicts and to safeguard common interests. Tangible achievements have included the comprehensive codification of ocean law, with the Law of the Sea Convention offering an authoritative framework. This is unique in the manner in which it links a number of more specialized global and regional agreements to its core structure.

The difficulty remains one of incorporating the ocean dimension into a number of pre-existing sectoral structures, where marine programmes have typically been grafted on to institutions focused primarily on terrestrial aspects, for example, marine science within UNESCO and fisheries within FAO. A notable exception to this general observation was the establishment of IMO, whose Assembly was convened for the first time in 1959. Conversely, and partly in response to this situation, more recent marine treaties have tended to produce independent action plans, strategies and – most importantly – funding mechanisms, carefully separated from broader existing structures. The result has been referred to as 'treaty congestion' – the proliferation of autonomous and semi-autonomous legal regimes and institutions, each with its own constituency among national government agencies, and net centrifugal tendencies in programming and financing. An example is waste management, where national and international mandates for both disposal at sea and ship-generated wastes conflict with those of land-based waste management and pollution control.

Institutional mechanisms for coordination and joint programming at the international level are notoriously weak, sometimes more symbolic than operational in nature. There have been encouraging signs of a more rational interagency division of work whenever new funding sources have become available and could be allocated on the basis of comparative advantage. A recent example is the GEF which, in addition to its three main 'implementing agencies', has used external executing agencies, such as IMO, for projects in the field of marine pollution control. However, less than 13% of overall GEF funding is currently allocated to International Waters projects, covering both marine and fresh waters. This bias is mainly due to stronger support by intergovernmental constituencies for other priorities, such as ozone depletion, climate change and biodiversity.

Efforts to improve this situation are confronted with numerous obstacles, including the following:

● There are still many countries where ocean information is scarce. What little there is may be stored in the head-quarters of international agencies and be difficult to obtain at the national level.

● Comparable and reliable information needs to be exchanged between neighbouring countries to facilitate the analysis of the problems of shared water resources – rivers or regional seas – as well as responses to them.

● Information needs to be organized in order to correlate scientific and environmental data, on the one hand, with social and economic findings, on the other, so as to identify cause/effect relationships, and hence the costs and benefits of different resource uses.

● The real value of marine and coastal resources needs to be incorporated into development choices so that inefficient resource use is discouraged and sustainability over the long term is not compromised. Indicators and methods of analysis for this purpose are still not widely available.

● Relevant information needs to be presented in a format useful to decision-makers. Many may not have the necessary aptitude or capacity to make full use of scientific, technical and statistical sources, while some scientists are unwilling or unable to recognize the

Institutional
mechanisms
for coordination
and joint
programming
at the
international
level are
notoriously
weak,
sometimes
more symbolic
than operational
in nature.

potential policy significance of the results of their work. Hence, the communication between producers and users of information needs to be enhanced.

● Finally, there are problems of public access to information. As noted in the previous chapter, major efforts are required to achieve greater transparency in respect of public disclosure of ocean-related information, not only from national authorities but also from international organizations.

Fragmentation at the national level

The structure of international ocean law and institutions essentially parallels that at the national level and reflects its weaknesses. National ministries and agencies responsible for fisheries and aquaculture, off-shore oil, ports and harbours and marine transportation, tourism, and the environment, all have an interest in ocean use, but the exercise of their mandates is rarely coordinated. Protection of the marine environment has not yet been subsumed into the sectoral concerns of national strategies, even in cases where the attainment of development goals is dependent upon the maintenance of environmental integrity. There are additional deficiencies in interministerial arrangements for addressing conflicts of use in coastal areas, with marine objectives often failing to receive the attention they deserve. The expansion of international legal instruments makes it increasingly difficult for governments and the general public to obtain and to maintain a clear picture of national objectives and commitments.

As new measures are adopted regularly at international meetings, giving more detailed content to the international legal framework, the burden grows. This hampers national ability to influence international policy and programme developments. Many countries find it difficult to elaborate more detailed rules and recommended practices to give effect to the multitude of international legal instruments in a mutually supportive manner. Fragmentation at the national level becomes more complicated due to the fact that many governments understandably allocate responsibilities for such matters as coastal zone management or waste disposal and sanitation to local authorities that, for a variety of reasons, may be unable to effectively discharge them. Therefore, in developing their positions, national authorities need to consult with local authorities, and to provide them with the assistance required to carry

> The expansion of international legal instruments makes it increasingly difficult for governments and the general public to obtain and to maintain a clear picture of national objectives and commitments.

out their responsibilities effectively. While this places additional demands on national officials, the pay-offs are considerable in the form of improved understanding at the local level of how national and international goals can be translated into meaningful local benefits.

THE UNFINISHED AGENDA OF OCEAN LAW

It is generally accepted that there is a need to further elaborate international rules:

• in the field of responsibility and liability for harm to the marine environment, especially for acts not prohibited by international law;
• updating the law of naval warfare, as outlined in Chapter 1;
• progressively transforming 'soft law' for ocean-related activities into treaties, and thus make it legally binding;
• incorporating the 'precautionary principle' into existing international agreements; and
• adopting detailed standards and recommended practices to strengthen global and regional framework agreements.

The top priority, however (as identified in paragraphs 8.15 and 39.8 of Agenda 21), remains the effective *implementation and enforcement* of the vast range of international ocean law now available. A number of instruments already in existence clearly suffer from underutilization. As a first step, states that have not become parties to global and regional treaties pertaining to their ocean-related activities should urgently do so – starting with the Law of the Sea Convention as the overarching instrument for global ocean governance. In so doing, they should refrain from the use of statements or reservations that weaken or fragment the unity of those agreements.

Many states have yet to give effect in law and practice, at the national and local levels, to commitments they have already entered into. They should be encouraged to take the necessary legislative and administrative action, to provide adequate financial support, and to promote motivation on the part of national officials, local authorities, individuals, and the private sector. In particular, proper implementation of the provisions of the Law of the Sea Convention should take into account the commitments emanating from other more recent agreements, such as the Convention on Biological Diversity, the Framework Convention on

Climate Change, and related objectives for national action specified in Chapter 17 of Agenda 21.

It follows that the first and most important item on the unfinished agenda of ocean law must be full compliance with existing treaty obligations. Concerted action by governments in this respect would be instrumental in defusing current widespread concern over: inadequate enforcement of international legal commitments; the development of appropriate instruments, including economic incentives; and the management of compliance. Lack of national compliance with treaty obligations, including the duty to report on legislative and administrative measures for implementation, has been well documented in the survey of existing international instruments carried out by the UNCED Preparatory Committee in 1991. Weak institutional arrangements at the national level often make it difficult to implement international law. Financial and technical assistance for *national capacity building* is therefore, essential to enable developing countries to participate effectively in agreements that are expected to yield international benefits. There is certainly scope for more organized and targeted efforts in ocean affairs to meet this concern.

The effectiveness of concerted national enforcement of international ocean law has been amply demonstrated by the 1982 *Paris Memorandum of Understanding* on Port State Control regarding compliance with IMO and ILO conventions, on vessel and crew safety and on marine pollution control, in Western Europe. Similar regional arrangements have been made for Latin America, the Asia–Pacific region, and the Caribbean. In effect, national authorities carry out enforcement functions on behalf of the international community simultaneously with their ordinary national enforcement tasks. The trend towards comparable arrangements for other ocean uses, such as fishing, is currently emerging.

Effective use should be made of available means established by the Convention for *dispute settlement*: the various arbitration processes, the International Tribunal for the Law of the Sea and the International Court of Justice; or, in more general terms, those means indicated in article 33 of the UN Charter. States should exchange pertinent information and engage in consultations at an early stage when there is a potential for conflict. In future agreements, provision should be made for effective means of dispute settlement and prevention. In the context of prevention, more attention needs to be paid to the innovative 'non-compliance procedures' set up under several recent environmental agreements, such as the 1990 amendments to the Montreal Ozone Protocol. With regard to territorial or resource disputes, the lessons of the Antarctic Treaty system should be followed, where contentious claims were effectively 'frozen' while the states involved developed joint management and conservation schemes for the areas and resources concerned. Other mechanisms to improve the utilization of existing instruments are available in the form of complaint procedures, some of the most innovative of which were described in the preceding chapter.

Law of the Sea Convention and the settlement of disputes

Part XV of the United Nations Convention on the Law of the Sea requires that States Parties to the Convention settle any dispute between them concerning the interpretation or application of the Convention by peaceful means in accordance with article 2, para. 3 of the Charter of the United Nations and shall seek a solution by the means indicated in article 33, para. 1 of the Charter. Where, however, no settlement has been reached, article 286 of the Convention stipulates that the dispute be submitted at the request of any party to the dispute to a court or tribunal having jurisdiction in this regard. Article 287 of the Convention defines those courts or tribunals as:

(i) The International Tribunal for the Law of the Sea (established in accordance with Annex VI of the Convention) including the Seabed Disputes Chamber;
(ii) The International Court of Justice;
(iii) An arbitral tribunal constituted in accordance with Annex VII of the Convention;
(iv) A special arbitral tribunal constituted in accordance with Annex VIII for one or more of the categories of disputes specified therein.

For the legal framework within the United Nations Convention on the Law of the Sea Procedures for settling seabed-related disputes, see Part XI – section 5, articles 186–191 – and Part XV. For non-binding procedures, see articles 279–285 and Annex V. For compulsory procedures entailing binding decisions, see article 287, Annexes VI, VII and VIII.

Source: UN Division for Ocean Affairs and the Law of the Sea (1997).

MAKING INTERNATIONAL AGREEMENTS WORK

Action to make ocean governance effective is conceivable at the following levels: national, regional, and global. In line with what is widely known as the 'subsidiarity' principle, action at the level closest to those directly concerned is normally most likely to succeed – national over regional or global – and should, by the same token, reach out to non-governmental organizations and local communities.

At the national level

Many countries at present lack the capacity to fully enjoy their benefits and to meet their obligations under the Law of the Sea Convention, with significant short-falls in legislation, institutions, instruments, expertise and financial resources. Particular problems are in evidence in the case of developing coastal states, which lack the capacity to undertake resource assessments in their extended national jurisdictions, to develop management systems and to monitor effectively the activities of other users. In such cases, exclusive economic rights may be more imaginary than real, and this will remain the case without more concerted efforts in the area of capacity building.

Capacity building will need to go hand-in-hand with education and training. Despite efforts that have been made in the fields of public education and awareness, there is still a general lack of perception of the importance of the oceans which, more often than not, is associated with a lack of political will to deal with marine and coastal issues. However, ocean governance in many countries is beset by technical and financial constraints. Even though the governments of many developing nations as well as countries in transition recognize the importance of the oceans for the economic well-being of their populations, they are not yet in a position to take advantage of marine potentials.

● Countries should establish at a high govern-mental level an appropriate policy and coordinating mechanism, to set and review national goals for ocean affairs.

There is obviously no single, optimal model for national institutions. A number of countries have set up special interministerial bodies to review ocean issues from an integrated perspective. Traditionally, these have

> Even though the governments of many developing nations as well as countries in transition recognize the importance of the oceans for the economic well-being of their populations, they are not yet in a position to take advantage of marine potentials.

concentrated on key economic sectors but, as sustainable resource use gains importance, they may take on broader objectives. For other countries, new institutional arrangements will be necessary to facilitate consideration of the full spectrum of ocean issues on a permanent basis.

The only sensible advice is for each country to establish an appropriate focal point – for example, an interministerial council or an office of ocean and coastal affairs – at the highest practicable administrative level, so as to ensure proper attention by all sectoral agencies concerned, including the authorities responsible for land-based and freshwater programmes and activities with likely negative impacts on marine and coastal areas. Its mandate should be to reconcile policy and programme discrepancies and to promote integration and sharing of ocean-related information. Its ultimate objective should be to guide and assist in the formulation of an integrated national policy for the oceans.

From the outset, a thorough review should be undertaken of the responsibilities of national departments and local authorities for ocean policies and programmes, and of the adequacy of information, and of technical and administrative capabilities, required to discharge them. This assessment would shed light on whether responsibilities are properly allocated and whether adjustments are required. It would also provide a basis for developing and updating a common national strategy to fulfil all ocean-related responsibilities. As a first stage, national strategies will concentrate on sectors vital to national goals and welfare. Specific targets and strategies provide a means for measuring progress and achievements. If well integrated, they also offer a means of assessing whether the policies and programmes supported in different international fora are mutually consistent.

In addition to the above high-level governmental mechanism, a national forum for ocean affairs would be desirable to ensure communication with experts, citizens' groups and stakeholders and the participation of civil society as a whole.

At the regional level

The shift toward an integrated approach at the national level should be reflected at sub-regional, regional and global levels, and appropriate linkages should be established between them. At the same time, it is essential to review the functions of global and regional agencies in order to assess their comparative advantages. Regional Commissions and networks – such as the

153

UNEP-sponsored Regional Seas Programme and IOC's regional subsidiary bodies – are essential when dealing with joint scientific research and integrated management, and potentially offer the most comprehensive legal and institutional framework for international cooperation in marine and coastal affairs. They also serve as vital links between ocean governance at the national and global levels.

The principal merit of regional arrangements is that they reflect the geographic scale of most problems of marine resources and ecosystems. While interregional diffusion of persistent organic pollutants has been demonstrated, and while a few marine species do circle the globe, most ecological linkages in the marine environment do not extend beyond regional seas. Regional approaches are appropriate for assessing the status of fisheries and other marine resources – on the basis of cumulative impacts on natural systems – and thus facilitate the identification of priorities for each region. Regional assessments can be structured to gauge progress in relation to both national goals and international conventions and programmes applicable in the region. They establish the foundation for more integrated approaches to interdependent conventions and institutions, for example, by relating the 1995 Programme of Action on Protection of the Marine Environment from Land-Based Activities to action under the Convention on Biological Diversity as well as several recent fishery agreements. They also help to reflect regional priorities in global decision-making. If states succeed in harmonizing goals and policies for the oceans at the regional level, these may eventually serve as a basis for global agreement.

● **Full advantage should be taken of regional organizations and programmes for the sustainable management of marine and coastal areas, including regional mechanisms for dispute settlement and prevention, and regional 'upwards harmonization' and implementation of standards.**

Regional approaches offer economies of scale for action specific to the region. They facilitate information exchange about problems, response options, and national ocean management strategies among countries with generally similar conditions. They allow a more direct involvement of the national constituencies and stakeholders concerned. In certain cases, they offer a promising framework for the settlement of disputes over localized territorial claims involving off-shore resources. However, concerted efforts are needed in some regions to strengthen human and institutional capabilities if the full potential of such cooperation is to be realized, especially (as pointed out in Chapter 2) in the context of regional systems for sustainable development and related marine science and technology. Furthermore, as illustrated by the Mediterranean Environmental Technical Assistance Programme (METAP), functional regional structures are more likely to attract external funding.

Regional approaches also encourage more open and meaningful discussions with experts and the public, and they allow decisions regarding the value of particular natural systems to be taken by those who use them. This necessarily involves local authorities in coastal areas, whether of mega-cities or of villages, as well as national government officials, representatives of international secretariats, resource users and non-governmental actors. Changes may be needed in the institutions established by, and associated with, regional agreements. A promising beginning has already been made in the Mediterranean, with the 1995–1996 revision of the Barcelona Convention and its Action Plan. Within this new framework, a Mediterranean Commission for Sustainable Development was established in which, for the first time, non-governmental representatives participate on an equal footing with governmental representatives.

The regional level is the obvious level at which to address some of the major new issues in the oceans. It is particularly suited for initiatives in the area of resource management, including broader-based approaches to the management of coastal zones and contiguous seas, as well as for initiatives in respect of technological cooperation and for programmes of technology transfer. It is also the most relevant level for joint surveillance and enforcement, and for initiatives aimed at linking regional seas programmes with agendas for peace and security. Initiatives in these and related areas are likely to acquire even greater importance in the twenty-first century.

Regional approaches also encourage more open and meaningful discussions with experts and the public, and they allow decisions regarding the value of particular natural systems to be taken by those who use them.

Mediterranean Commission for Sustainable Development

An important initiative taken in the wake of UNCED is the creation of the Mediterranean Commission for Sustainable Development, which met for the first time in Rabat, Morocco in 1996. The Commission's function is to advise the States Parties to the 1976 Barcelona Convention on regional issues involving interactions between environmental and socio-economic obligations as well as to monitor progress on sustainable development. Thus, for example, the Commission is charged with examining the environmental impact of the policies associated with the creation of the new Euro-Mediterranean trading area, as well as the interrelated environmental and economic effects of the development of tourism in the region.

The Commission has two important innovative characteristics: it includes official representation from ministries of development and/or economic affairs, and not just from environmental administrations, thus conferring a multi-sectoral and integrated approach to the management of the Mediterranean. The second feature is that it has two other segments, apart from the central government; it has a segment elected from representatives of municipal authorities, and local constituted bodies (e.g. chambers of commerce, trade unions, trade organizations) with particular reference to coastal communities; and the other elected segment represents NGOs. The most remarkable aspect is that all the three segments are on a par with voting rights.

The Commission operates through the UN Mediterranean Action Plan, based in Athens (Greece), of the Barcelona Convention. The first regional entity of its kind, the Commission collaborates with the UN Commission on Sustainable Development (New York).

Source: Salvino Busuttil, former Coordinator, Mediterranean Action Plan (1998).

At the global level

At the global level, the Law of the Sea Convention, the central regime for ocean governance, has established a new treaty system of ocean institutions. Under the umbrella of the Convention, a number of 'sub-regimes' can be identified, each of which deals with specialized matters. The most important of these sub-regimes cover:

● The sustainable management of marine living resources, the focus for which is FAO, including its network of regional fisheries commissions and conventions;

● Shipping and marine pollution control, centered on IMO and several related convention-based institutions;

● The marine environment, the main responsibility for which has been assigned to UNEP, including its network of regional seas agreements and action plans;

● Marine scientific research and associated ocean services and management, centered on IOC;

● Deep seabed mineral development, through the International Seabed Authority.

The Law of the Sea Convention has thus provided a comprehensive framework that has given the UN system its pivotal role in ocean governance. With all the shortcomings noted previously, this combination of regimes will continue to form the basis for ocean governance for the foreseeable future – not least because all component 'sub-regimes' have autonomous or semi-autonomous governing bodies, each with its own constituency of member states or participating states.

Not surprisingly, a workshop jointly sponsored in 1995 by the Brazilian and British governments concluded that 'there is no need to create any new global agency or institution for taking decisions at the global level on questions affecting the marine environment'. However, an overview mechanism in the framework of the UN could help ensure that potential inconsistencies in legal instruments evolving at global and regional levels are brought to the early attention of governments, and that plans to develop new instruments do not contradict or duplicate each other. Without an overview of

Institutionally,
the General
Assembly
remains the
forum that is
competent to
consider in an
integrated
manner
developments
related to the
Law of the Sea
and ocean
affairs...

international programmes and priorities identified in each region, it is not possible to make fair assessments regarding the allocation of scarce international resources, or to identify the areas in which joint efforts would be most cost-effective.

Since the adoption of the Convention in 1982, the UN General Assembly has reviewed annually all aspects of ocean affairs and the Convention, based on a report prepared by the Secretary General. The documentation is generally of high quality, but its one-day discussion is insufficient for a thorough review of developments in the oceans. Few governments are in a position to prepare adequately for the meeting or to send ocean experts to take part in it. The Commission on Sustainable Development (CSD) has since 1993 reviewed the forty chapters of Agenda 21 in a five-year cycle, and will reconsider the ocean chapter in 1999. Following the entry into force of the Convention in 1994, a series of meetings of the States Parties to the Law of the Sea Convention (SPLOS) has been held. So far, the meetings have been mainly confined to taking the decisions required by the Convention, such as the adoption of a budget and elections to the International Tribunal for the Law of the Sea and the Commission on the Continental Shelf.

Institutionally, the General Assembly remains the forum that is competent to consider in an integrated manner developments related to the Law of the Sea and ocean affairs, but there is clearly a need to strengthen the on-going review of ocean affairs. This could be achieved through consideration by a Committee of the Whole of the General Assembly, or by expanding and formalizing the existing, but informal, consultations preceding the General Assembly debate. In each case, preparations should make use of CSD efforts to integrate national reports, but would also need to go beyond them to cover a broader range of information, policy and operational matters. Inputs from regional assessments and from regional discussions of ecologically linked conventions would be vital. Provision could also be made for dialogue with specialists in ocean science, economics, sociology and other fields.

While the review meetings would essentially remain high-level intergovernmental events within the UN framework, contributions from, and attendance by, other competent intergovernmental and non-governmental organizations should be encouraged. In this respect, special attention should be given to competent international scientific bodies, such as ICES and ICSU.

Additionally, it has been suggested that the meetings of SPLOS could serve as an appropriate forum for the discussion of ocean issues. Although not originally seen as a formal institution, SPLOS meetings are beginning to emerge as a potential forum for the regular review of major marine policy matters. In order to facilitate subsequent deliberations in the General Assembly – and without affecting the mandate of CSD – the scope of the review would have to cover all developments under the Convention and its implementing agreements. For this purpose, however, the mandate of SPLOS would have to be modified, and provision possibly made for participation by parties to ocean-related conventions, such as those on biodiversity and climate change, which are not party to the Convention, and for observer participation by other relevant treaty bodies as well as by non-governmental ocean interests.

In his 1997 report *Renewing the United Nations: A Programme for Reform*, the Secretary-General advanced another idea – that of giving the UN Trusteeship Council a role to play in the review of ocean affairs. The Council (which at present consists only of the permanent members of the Security Council) can hardly be said to reflect the democratic–participatory approach to ocean governance advocated by this report. It would have to be reconstituted, which presumably requires a revision of the UN Charter.

● **The discussion of ocean affairs within existing fora of the UN system should be strenghtened and supplemented by a comprehensive review of the mandates and programmes of all UN bodies and agencies competent in ocean affairs.**

● **To advance the process of change and innovation within the intergovernmental system, consideration should be given to convening, at an early opportunity, a United Nations Conference on Ocean Affairs.**

This Conference would aim at placing the oceans prominently on the international and national political agendas, in order to facilitate making the present system of ocean governance more coherent, responsive and democratic. The proposed conference would not be law making. It would take as its foundation the Law of the Sea Convention as well as other relevant international treaties and programmes. The conference would draw upon a review of all salient issues relating to the use and abuse of the oceans and of ocean space in their dynamic interactions with coastal areas, rivers and land-based activities.

SPEAKING UP FOR THE OCEANS

The recommendations set out above are essentially directed at governments in recognition of the key role they must play in shaping a more effective system of ocean governance. They are presented, quite deliberately, with an eye to political feasibility and, together, they should establish a solid foundation for making the present system more coherent, more responsive and more democratic.

However, beyond these recommendations, this report argues that the challenges posed by the oceans will not, and cannot, be adequately met unless global civil society is provided with significantly enlarged opportunities to participate in ocean affairs and to exert its influence on processes of change. Change and innovation within the governmental and intergovernmental systems can also be facilitated by initiatives taken outside these systems. The plea for such initiatives should not be regarded as a plea for the creation of institutions that would exist in competition with governmental bodies or which would seek to duplicate their work. Rather, they should be interpreted as complementary measures that seek to make the system of ocean governance more democratic. The development of new systems of ocean governance must thus be defined in ways that foster the participation and involvement of stakeholders and enable them to speak up for the oceans.

Here, it is possible to travel new paths set out by the 1992 Rio Earth Summit, which reiterates the importance of participation, and by the Law of the Sea Convention's implementing agreement on straddling fish stocks that makes such participation possible. However, it must be recognized that efforts to enlarge participation will take place in an institutional environment that, for historical reasons, is strongly delineated along sectoral lines. While sectorally based participation should continue, the Commission recognizes the need for new initiatives that are not confined to traditional sectors, and which enlarge opportunities for civil society to participate directly in ocean affairs.

The Commission considers it to be essential to ensure independent monitoring (to enhance *transparency*) and independent assessment (to enhance *accountability*) of ocean affairs. Accordingly, it recommends:

> The development of new systems of ocean governance must thus be defined in ways that foster the participation and involvement of stakeholders and enable them to speak up for the oceans.

● **The establishment of a World Ocean Affairs Observatory in order to independently monitor the system of ocean governance and to exercise, on a continuous basis, an external watch on ocean affairs.**

In the first instance, the Observatory would serve as a focal point for bringing together relevant information from other sources – official and unofficial, including intergovernmental, governmental and non-governmental institutions or networks. The information so obtained would be used by the Observatory to produce periodic 'state of the oceans' reports as well as ad hoc studies of urgent ocean issues. At the same time, the Observatory would serve as an interactive 'virtual' observation site for all ocean-related information in the World Wide Web, providing direct electronic links to all relevant (public and private) Internet sites.

Today, there are a number of precedents for successful performance of this monitoring role – in the fields of human rights, environment and disarmament – by non-governmental bodies such as Amnesty International, Greenpeace and SIPRI. These independent bodies play a crucial 'watchdog' role, and hence may serve as precedents for the work of the Observatory.

As a complementary measure, the convening of an *Independent World Ocean Forum* would allow public assessments by an independent assembly representing civil society and all stakeholders. It would further allow the actors to be held accountable for the use of ocean space and for the management of its resources. This objective could best be served through a broad-based Forum, with a mandate to undertake a comprehensive review of current ocean issues every three or four years. The Forum would not be part of the intergovernmental structures previously discussed; it would have no decision-making powers; and it would operate as a 'recurrent event' rather than as a permanent institution. The Forum would draw on studies and other outputs generated by the Observatory, and on interactive electronic communications with open public participation during pre-sessional and intersessional periods. Another complementary measure would be the appointment of an independent Ocean Guardian, as suggested in the previous chapter.

Such initiatives would enable those with an interest in the oceans – and their interactions with land-based activities, rivers, and coastal areas – to better articulate their concerns and express their aspirations. They would empower new voices to speak up for the oceans.

Annexes

A. Some Salient Facts about the Oceans

B. Regional or National Inputs

C. Selected Sources

D. Acronyms and Abbreviations

E. The Commission and its Work

OCEAN FACTS

Annex
A
1
Basic
Facts

71% of the
earth's
surface is
ocean

- The oceans cover approximately 361 million sq. km (139 million sq. mi), or 71% of the earth's surface.

- The Northern Hemisphere is 60.7% sea and 39.3% land; the Southern Hemisphere is 80.9% sea and 19.1% land.

- The oceans' total volume is approximately 1348 million cu. km (324 million cu. mi).

- The average depth of the oceans is 3733 m (12 247 ft) and the deepest point is 11 022 m (36 163 ft).

- At the deepest point, the pressure of the ocean is over 1 tonne per square centimetre (8 tonnes per sq. in), the equivalent of one person trying to hold up 50 jumbo jets.

- The average temperature of the oceans is 3.9 °C (39 °F), but it ranges from below freezing under the ice shelves of Antarctica to as high as 37 °C (98.6 °F) in the Arabian Gulf.

- Of all volcanic activity, 90% occurs in the oceans. In 1993, scientists located the largest known concentration of active volcanoes on the seafloor – in an area of the South Pacific the size of New York State – which hosts over 1100 volcanic cones and sea mounts.

- The longest mountain range in the world is the Mid-Ocean Ridge. The ridge is approximately 64 000 km (40 000 mi) long, and, on average, more than 2000 km (1200 mi) wide.
Circling the globe from the Arctic Ocean to the Atlantic Ocean, passing into the Indian Ocean and crossing into the Pacific, it is four times longer than the Andes, Rocky Mountains and Himalayas combined.

THE WORLD'S WATER

The oceans contain by far the most water on the planet. The percentage of water found in the oceans, and elsewhere:

	%
The ocean	>97
Ice on land	1.9
Groundwater	0.5
Rivers and lakes	0.02
Water in the atmosphere	0.001

Source:
Garrison, T. (1996).

THE MAJOR OCEANS

Approximate sizes of the three ocean basins:

	Area (million sq. km/miles)	% of ocean/ earth's surface
Pacific Ocean	180 /70	50.0/35.5
Atlantic Ocean	107 /42	29.4/20.8
Indian Ocean	74 /29	20.6/14.5

	Average depth (m/ft)	Greatest depth (m/ft)
Pacific Ocean	3940/13000	11022/36163
Atlantic Ocean	3310/10920	8605/28231
Indian Ocean	3840/12670	7258/23812

Many popular texts give smaller areas for the three main oceans, choosing instead also to list smaller seas such as the Sea of Okhotsk, Bering Sea and Arctic Ocean. However, difficulties with delineating boundaries between, for example, the Pacific Ocean and the South China Sea leads to imprecision and inconsistencies. Accordingly, oceanographers prefer to use the more accurate and precise measurements given above, in which the figures for each ocean basin include those for adjacent seas.

Source:
Gross, M. G. (1995).

THE HYDROLOGICAL CYCLE

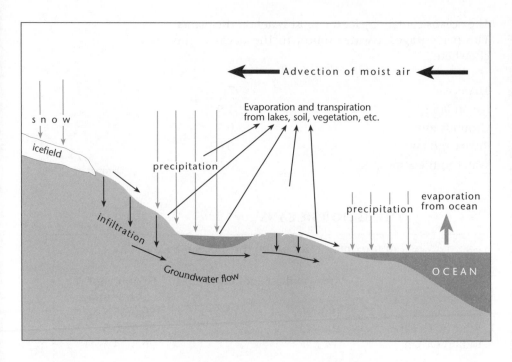

← Advection of moist air ←

Evaporation and transpiration
from lakes, soil, vegetation, etc.

s n o w

icefield

precipitation

precipitation

evaporation
from ocean

infiltration

Groundwater flow

OCEAN

Source:
Harvey, J. G. (1979).

Every year, around 505 000 cu. km (121 200 cu. mi) of seawater evaporate from the surface of the earth, entering the atmosphere as water vapour.

Of this, 430 000 cu. km (103 200 cu. mi) come from the ocean. However, precipitation over the ocean is only 390 000 cu. km (93.6 cu. mi) per year. The excess evaporation from the ocean falls as rain, sleet and snow on the surface of the land.

(a)

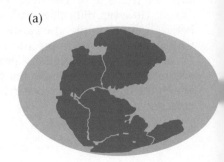

ORIGINS

There is but a finite amount of water on the planet, and most of it has been on earth since very early in the planet's geological history.

The ocean formed gradually by separation of water and gas from silicate rocks. Initially, the water existed mainly in the form of water vapour in the atmosphere.

As the earth cooled, probably around 3800 million years ago, this water vapour fell as rain, forming rivers and filling lower-lying basins to create the ocean.

The present ocean basins are relatively young. Approximately 225 million years ago, there was just one single continental land mass, Pangaea, which was surrounded by a giant ocean, Panthalassa. This began to break up, about 180 million years ago, into two smaller land masses Laurasia and Gondwanaland.

Not until about 20 million years ago did the size and shape of the continents and the ocean look like the world map with which we are familiar.

They are continuing to change and shift today, a combination of seafloor spreading and plate tectonics causing the Atlantic to widen by about an inch each year, while the Pacific shrinks.

Below: Movements of lithospheric plates during the past 225 million years following (a) the break-up of Pangea, (b) formation of the Atlantic, and (c) opening of the Indian Ocean. *Source*: Gross, M. G. (1995).

(b)

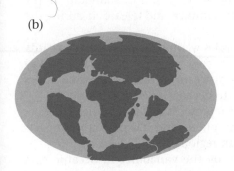

(c)

THE OCEAN AND CLIMATE

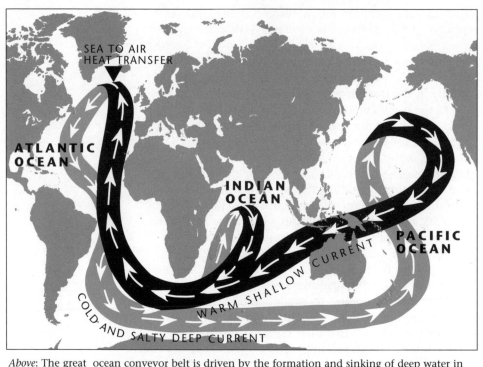

SEA TO AIR
HEAT TRANSFER

ATLANTIC
OCEAN

INDIAN
OCEAN

PACIFIC
OCEAN

WARM SHALLOW CURRENT

COLD AND SALTY DEEP CURRENT

Above: The great ocean conveyor belt is driven by the formation and sinking of deep water in the Norwegian Sea. *Source*: Broecker, W. S. (1991).

It takes about 3200 times more heat to warm a given volume of water by 1 degree as it does to warm the same volume of air.

Thus the ocean has been described as the flywheel of the climate system because of its huge heat storage capacity. It stores energy when it is in abundant supply during the day, or in summer, and releases it during the night, or in winter.

The ocean also carries warm equatorial water towards the poles and releases cold water to the equator by way of surface and deep ocean currents over time scales ranging from years to decades or even centuries.

This movement represents a source of heat transfer as great as the atmosphere, and has profound impacts for the climate, both regionally and globally.

El Niño, caused by the variability in oceanic heat transfer, has recently become one of the most publicized examples of the ocean's impact on the global climate.

NUMBERS AND DIVERSITY

Estimates vary widely as to how many species there are in the ocean, and how the number of marine species compares with the number that live in terrestrial environments.

It is generally considered that there are fewer species in the sea than on land (primarily because of the vast numbers of insect species), but there is much uncertainty.

One 1993 review concluded that there were 178 000 marine species in 34 phyla; by contrast, a different study suggested that deep seabed communities alone, until relatively recently considered comparatively void of life, may contain as many as ten million species, more than have been identified on land. However, this theory is controversial and has been disputed by other scientists.

At the level of phyla, the broadest category of classification after kingdoms, life in the marine environment appears to be more diverse than that on land.

There are 43 marine phyla and only 28 terrestrial ones; according to the *Global Biodiversity Assessment*, produced by the United Nations Environment Programme, of 33 extant phyla of animals on the planet, 32 are found in the ocean, and 15 are exclusively marine.

2
Life
in
the
Oceans

There are 43 marine phyla and only 28 terrestrial ones

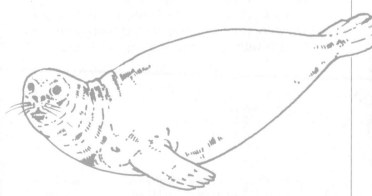

The Mediterranean monk seal, *Monachus monachus,* one of the largest seal species, has been classified as endangered since 1966.
Source: Cousteau, J. (1996).

PLANKTON

The basis of the marine food web is a community of microscopic organisms called *plankton*. With very few exceptions, marine communities all depend on the production of *phytoplankton* and *zooplankton* for their sources of energy.

Phytoplankton are tiny, normally single-celled, algae, the smallest less than 2 microns (a hair's width) thick, and the largest over 2 mm (0.078 in) long. There are two main types: diatoms, which are generally dominant in temperate and high latitudes; and dinoflagellates, which tend to occur more frequently in warmer waters.

Distribution and abundance of phytoplankton communities is determined by a number of variables, including light, temperature and nutrient levels in the water. In general terms, there is little plankton production in tropical regions, largely because of a lack of nutrients.

Because nutrient levels are generally higher in coastal regions, the areas of highest phytoplankton concentrations, and hence the most productive areas of the world's oceans, tend to be in coastal waters.

Phytoplankton grow and multiply so quickly that, if they were not grazed by zooplankton, they would double in quantity within one or two days.

Zooplankton are animals which range in size from microscopic protozoans to the mega-zooplankton, which can attain over 2 mm (0.078 in) in length. They are generally herbivorous, grazing on the phytoplankton, although some eat other zooplankton.

The most abundant zooplankton are copepods, small crustaceans which sweep the phytoplankton into their mouths with constantly moving front limbs.

Left: This shell of a *silicoflagellate* is 0.03 mm (0.0012 in) in diameter, including spines. As the name suggests, these tiny phytoplankton have silica skeletons and are widespread in colder seas worldwide. *Source*: Gross, M. G. 1995.

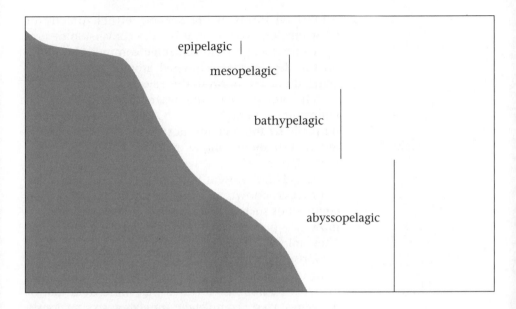

Epipelagic zone. Extends from the surface down to a depth of about 100 m (328 ft). The lower boundary marks the limit of the penetration of sufficient sunlight to support photosynthesis. Almost all the world's fisheries resources live in this zone.

Source:
Waller, G., ed. (1996).

Mesopelagic zone. Extends down to about 1000 m (3281 ft), the limit of light penetration. Many of the creatures living in this zone make daily vertical migrations. The food supply is either transported by these migrating animals or sinks in the form of detritus from the surface.

Bathypelagic zone. The total amount of life and the number of species is relatively small compared to the layer above. In this dark realm, species such as angler fish use light organs to lure their prey.

Abyssopelagic zone. Extends from about 2000 m (6562 ft) downwards. Life becomes increasingly sparse until about 100 m (328 ft) off the seafloor, where it once again becomes more prevalent.

DEEP-SEA VENTS

Archaea
may have
been the
very first
forms of
life on the
planet

Almost all forms of life depend, either directly or indirectly, on photosynthesis – the conversion of sunlight into energy. However, some communities living on the deep seabed, clustered around hydrothermal vents, do not conform to this rule.

The vents heat the water around them to temperatures as high as 113 °C (235 °F), temperatures normally far too high for most forms of life to survive. But they harbour unique microbes known as hyperthermophiles, which thrive in these hot temperatures.

Instead of photosynthesis, these microbes survive by a process known as chemosynthesis, in which compounds such as hydrogen sulphide – poisonous to most life – provide the hyperthermophiles with energy. These microbes then provide the basis for a food chain comprising a variety of equally unique lifeforms.

In 1996, analysis of the DNA structure of one species of hyperthermophile revealed that more than half of its genes were completely unrelated to previously known genes.

As a result, scientists believe that these species of chemosynthetic microbes should be classified in their own kingdom, alongside the two into which living things are presently classified.

This third kingdom has been called *Archaea*, or 'ancient ones', because it is believed that these species, or their direct ancestors, may have been the very first forms of life on the planet.

THE TEN COUNTRIES WITH THE MOST COASTLINE

	kilometres	miles
Canada	243 791	151 484
Indonesia	54 716	33 999
Greenland	44 087	27 394
Russia	37 653	23 396
Philippines	36 289	22 549
Australia	25 760	16 006
Norway	21 925	13 624
United States	19 924	12 380
New Zealand	15 134	9 404
China	14 500	9 010

3

Coasts, Islands and Populations

Source:
United States National Imaging and Mapping Agency (NIMA).

THE OCEANS' TEN LARGEST ISLANDS

Island, location	sq. km	sq. mi
Greenland, NW Atlantic	2 175 600	848 500
New Guinea, SW Pacific	792 500	309 100
Borneo, W Mid-Pacific	725 500	283 000
Madagascar, Indian Ocean	587 000	229 000
Baffin NW, Atlantic	507 500	198 000
Sumatra, Indian Ocean	427 300	166 700
Honshu, Sea of Japan	227 400	88 700
Great Britain, NE Atlantic	218 100	85 100
Victoria, Arctic Ocean	217 300	84 800
Ellesmere, Arctic Ocean	196 200	76 500

Source:
National Geographic Atlas of the World. (1996).

COASTAL MARINE ENVIRONMENTS

Ecologists divide coastal regions into three principal zones: the *supralittoral*, at the edges of the reach of sea spray; the *eulittoral*, the lower boundary of which generally lies at, or a little above, the mean low water of spring tides; and the *sublittoral*, which extends from the lower limit of the eulittoral to the point where macroalgae (or seaweed) growth no longer occurs.

Each of these zones is generally further divided into sub-zones, each containing a variety of habitats with their own diverse communities of organisms. Some of these habitats include:

● **Shoreline environments.** The areas just above the high tideline, including the strandline, sea cliffs and sand dunes. The shoreline provides many diverse breeding sites for seabirds, shorebirds and terrestrial animals.

● **Intertidal environments.** The areas between the high and low tide marks. The animal and plant life varies widely, depending on, among other conditions, whether the shore is rocky, muddy or sandy.

● **Estuaries and saltmarshes.** Estuaries are semi-enclosed, coastal areas of water in which the ocean water is significantly diluted by fresh-water run-off. Saltmarshes and adjacent mudflats are important breeding and feeding grounds for seabirds, shorebirds and waterfowl.

● **Mangroves**. Mangroves replace saltmarshes in tropical and subtropical regions. They consist of a number of trees and tree-like shrubs that are tolerant of both saltwater and brackish water conditions. They provide a number of different habitats, which shelter a wide variety of wildlife.

● **Kelp forests.** These are found in cold temperate regions below 20 °C (68 °F). They start at the low water mark and extend into sub-tidal areas. Kelp is the name for large brown algae, up to 50 m (165 ft) in length, which have their own associated plant and animal communities.

● **Seagrass meadows.** Seagrasses are found in clear, sheltered waters over a wide range of latitudes world-wide. The most abundant areas of seagrass are found just below the tidal level, and can occur down to a depth of about 50 m (165 ft). Typically, they are found in enclosed and sheltered bays, lagoons and offshore barriers.

● **Coral reefs.** Highly diverse habitats found in shallow, tropical waters where the average sea temperature is above 18 °C (64.4 °F).

COASTAL CITIES AND POPULATION

Estimates vary as to the number of people world-wide who live near the coast. One widely quoted figure is that as much as about two-thirds of the global population lives within 100 km (62 mi) of the shore. However, a 1997 study revised these figures downward.

Using a public digital map of the world human population, and the World Vector Shoreline, produced by the US National Imagery and Mapping Agency, the study concluded that 37% (2.1 billion) of the 1994 population (5.6 billion) lived within 100 km (62 mi) of a coastline, and approximately 44% (2.5 billion) within 150 km (93 mi).

Coastal populations are increasing at a higher relative rate than the population in general. In the United States over the period 1960–1990, the population in coastal counties grew by 41 million, an increase of 43%. Between 1983 and 1991, 90% of all building activity in Australia took place within the coastal zone.

Of cities with populations above 2.5 million inhabitants, 65% are located along the coast. Twelve of the twenty most highly populated urban areas in the world are located within 160 km (100 mi) of the coast.

Twelve of the twenty most highly populated urban areas in the world are located within 160 km (100 mi) of the coast

THE TEN LARGEST URBAN AREAS
within 160 km (100 mi) of the coast

Urban Area	Country	No. of people in millions
Tokyo-Yokohama	Japan	32.3
New York	USA	19.7
São Paulo	Brazil	15.2
Los Angeles	USA	15.0
Shanghai	China	12.9
Bombay	India	12.6
Buenos Aires	Argentina	11.0
Calcutta	India	10.9
Beijing	China	10.8
Seoul	S. Korea	10.6

Source:
National Geographic Atlas of the World. (1996);
Population Reference Bureau, Washington, D.C.

COASTLINES AT RISK

In coastal regions around the world, population and development pressure, as well as human activities further inland, are combining to degrade and destroy marine habitat, threaten marine wildlife and alter coastal marine ecosystems.

Israel, for example, is literally running out of sand as a result of decades of poorly regulated sand mining for construction. The mining has destroyed habitat for the animals and plants which live in the dunes along Israel's southern Mediterranean coast, and the coastal highway has cut off wildlife from the hinterland, leaving them with nowhere to go as the coastal sand disappears.

The discharge of nutrients and pollutants into the sea in some coastal areas has led to the disappearance or near-disappearance of some species, an increasing frequency of phytoplankton blooms, and the growth of so-called 'dead zones' where diminished oxygen supplies prevent the growth of any marine life.

- The anoxic 'dead zone' in the Gulf of Mexico lasts approximately eight months a year and can extend over 9000 sq. km (3510 sq. mi).

- In the Baltic Sea, a four-fold increase in nitrogen and an eight-fold increase in phosphorous input since the period before 1900 has led to the larger, bottom-dwelling animal life becoming almost extinct over an area of about 70 000 sq. km (27 300 sq. mi).

- According to the World Conservation Union, as much as 10% of the world's coral reefs have been degraded beyond recovery, and another 30% is likely to decline by the end of the first decade of the twenty-first century.

  Those at greatest risk are in South and South-East Asia, East Africa and the Caribbean, but the problem is widespread: of 109 countries where reefs are known to occur, significant reef degradation has occurred in 93.

- At its peak, harvesting and destruction of mangroves worldwide was estimated at 1 million hectares (2.5 million acres) per year. Causes of such destruction included conversion for urban and industrial development and agricultural cultivation.

- From 1983 to 1994, the number of published scientific reports of seagrass loss as a result of human activities multiplied almost five-fold over the previous decade. More than 90 000 hectares (223 000 acres) were reported as being lost, although the actual total was certainly greater.

As much as 10% of the world's coral reefs have been degraded beyond recovery

COASTS, ISLANDS AND CLIMATE CHANGE

Over the past 100 years, global sea-level has risen by between 10 and 25 cm (4–10 in); according to the Intergovernmental Panel on Climate Change (IPCC), 'it is likely that much of the rise in sea-level has been related to the concurrent rise in the global temperature'.

Although uncertainties remain, the IPCC speculates that warming and subsequent expansion of the oceans may account for 2 to 7 cm (0.8–2.8 in) of this rise, while the melting of glaciers and sea-ice may have contributed a further 2 to 5 cm (0.8–2.0 in). The remainder is principaly caused by land subsidence.

In the absence of any remedial measures, the IPCC estimates that total average sea-level rise is likely to be about 12 cm (4.7 in) by 2030 and 49 cm (19.3 in) by 2100. This increase would not be uniform, with the greatest increases expected in the North Atlantic, while in some areas, such as the Ross Sea in Antarctica, sea-levels might actually fall.

For low-lying coastal environments, the impacts of such sea-level rise will likely be profound. In Bangladesh, for example, about 7% of the country's habitable land (with about 6 million population) is less than 1 m (3.3 ft) above sea level, and about 25% (with about 30 million population) is below the 3 m (10 ft) contour.

In this region, it is estimated that the combination of thermal expansion of the oceans and land subsidence, among other factors, will lead to sea-level rise of about 1 m (3.3 ft) by 2050 and 2 m (6.6 ft) by 2100, threatening large areas of habitable and agricultural land essential to the well-being of the population.

At the same time, rising sea-levels, particularly when combined with the destruction of coastal mangroves, will significantly increase the region's vulnerability to storms, and lead to greater saltwater intrusion into fresh groundwater resources.

Similarly, in the Nile delta region of Egypt, about 12% of the country's arable land, with a population of 7 million, would be affected by a 1 m (3.3 ft) rise in sea level. A sea-level rise of just 50 cm (1.65 ft) would flood an area of 40 000 sq. km (15 600 sq. mi) – about the area of the Netherlands – along the eastern coastline of China.

Perhaps at greatest risk of all are the inhabitants of small islands and coral atolls, such as the Maldives or Marshall Islands in the Pacific. A sea-level rise of just 50 cm (1.65 ft) will significantly reduce their area and contaminate up to 50% of their groundwater. The necessary protective actions are beyond the resources of such small nations.

Over the past 100 years, global sea-level has risen by between 10 and 25 cm (4–10 in)

4

Ocean Exploration and Discovery

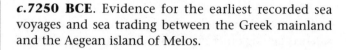

c.7250 BCE. Evidence for the earliest recorded sea voyages and sea trading between the Greek mainland and the Aegean island of Melos.

c.4000 BCE. Ancient Egyptians develop ship-building and ocean pilot skills. Although their voyages may have been confined to the Mediterranean, it has been suggested that they might have made their way across the Atlantic and thus contributed to the settling and peopling of the Americas.

c.2500 BCE. The inhabitants of New Guinea begin a process of seaborne migration into the Pacific.

c.1500 BCE. The Phoenicians, a people from a narrow corridor of land along the coast of modern-day Syria and Lebanon, set sail on voyages of exploration and trade. Over the next few centuries, they explore the whole of the Mediterranean Sea, reach Britain and Spain, where they mine silver, and possibly discover the Azores, 965 km (600 mi) west of Portugal.

1492 BCE. At the instigation of Queen Hatshepsut, an Egyptian expedition sails south through the Red Sea to the land of Punt (presumably the southern shore of the Gulf of Aden, or the Somali coast of Africa) to trade – a round trip by sea of over 5000 km (3000 mi).

600 BCE. According to the Greek historian Herodotus, a Phoenician crew, under the orders of Egyptian pharaoh Necho, circum-navigates Africa from east to west, re-entering the Mediterranean through the Strait of Gibraltar.

479 BCE. Using ancient nautical knowledge, the Greeks defeat the Persians at the Battle of Salamis. This battle allows the Greeks to become the dominant culture of the time, and lays the foundation of western civilization as we know it.

218 BCE. Chinese sea captain Hsu Fu sets sail in search of the elixir of immortality, possibly reaching North America.

c.100 BCE. Arab dhows regularly sail from Red Sea ports to the west coast of India.

54 BCE–AD30. The Roman, Seneca, puts forward his hypothesis on the 'hydrological cycle'.

150 The Egyptian geographer, Ptolemy, divides the Earth into 360 degrees of latitude and longitude.

200–900. The Polynesians spread out from Samoa to colonize much of the Pacific.

414. After 15 years exploring India, Fu-Hsien, a Chinese Buddhist monk, returns to China by undertaking a long sea voyage via Sri Lanka and Djawa.

673–735. The English monk, Bede, describes lunar influences on the tides.

982. The Norseman, Eric the Red, lands on Baffin Island.

1325–48. Ibn Battuta sets out on an epic journey from Tangier, including sea journeys to Somalia, Tanzania, India, Sumatra and China.

1405–33. A series of seven Chinese expeditions set out for the Indian Ocean. Three reach India, three the Persian Gulf and one the east coast of Africa.

1418. Prince Henry the Navigator of Portugal promotes the study of navigation and sends ships out to explore the world. The age of European exploration begins.

1492. Christopher Columbus lands in America.

1497–98. Portuguese explorer Vasco da Gama makes a voyage around Africa to India, opening a new trade route.

1519–22. A Spanish expedition under the command of Ferdinand Magellan becomes the first to circumnavigate the world.

1675. The Royal Observatory is established in Greenwich, England, and sets the line of longitude at 0° (the Greenwich Meridian).

1725. Luigi Marsigli publishes *Histoire Physique de la Mer*, the first modern treatise on oceanography.

1768–79. Captain James Cook undertakes three major voyages of discovery and exploration in the Pacific, and becomes the first to cross south of the Antarctic Circle.

1831–36. The voyage of the *Beagle,* during which the ship's naturalist, Charles Darwin, makes observations that provide the basis for his theory of evolution.

1839–43. Scientists on the *Erebus* and *Terror*, under the command of Sir James Clark Ross, discover life at a depth of 732 m (2400 ft) in the Antarctic.

1855. The American, Matthew Fontaine Maury publishes *The Physical Geography of the Seas.* Immensely popular, the book is translated into several languages; Maury becomes known as the 'father of oceanography'.

1858. After years of seabed surveys, the first telegraph cable is laid.

1872–76. The *Challenger* expedition, the first great voyage of oceanographic exploration. This 41-month circum-navigation of the world visits all the world's oceans, except the Arctic, discovering more than 700 new genera and 4000 new species, revealing the nature of the seabed, and collecting data for the further study of oceanic circulation.

1892–95. The Norwegian explorer, Fridtjof Nansen, allows his ship, *Fram*, to drift, trapped, in the Arctic ice. He confirms that there is no Arctic continent, and the meteorological and oceanographic data he collects lead to a much more sophisticated understanding of the complex hydrography of Arctic waters.

1895–98. The American, Joshua Slocum, becomes the first man to circum-navigate the world single-handed, in his 11-tonne sloop *Spray*.

1902. V. Walfrid Ekman develops a mathematical explanation (Ekman's Spiral) linking wind direction and current flow.

1912. The German scientist, Alfred Wegener, proposes his theory of 'continental drift'.

1920. Alexander Behm bounces sound waves off the bottom of the North Sea, advancing the development of 'echo-sounding'.

1934. Zoologists William Beebe and Otis Barton descend to a depth of 923 m (3028 ft) in a tethered bathysphere, becoming the first people to observe deep-sea life.

1943. Jacques-Yves Cousteau and Emile Gagnan develop the self-contained underwater breathing apparatus, or SCUBA.

1947. Thor Heyerdahl and five others sail 6400 km (4000 mi) across the Pacific from Peru to Tahiti on their raft *Kon Tiki*, proving that early Polynesians could have made such a voyage.

1951. Using echo-sounding, the British ship *Challenger II* finds the deepest chasm in the ocean, 11 km (7 mi) beneath the surface near Guam. It is subsequently named the Challenger Deep.

1953. Auguste Piccard and his son Jacques enter their bathyscaph *Trieste*, and descend to a depth of nearly 3.2 km (2 mi).

1958. The nuclear-powered *USS Nautilus* becomes the first submarine to pass under the Arctic icecap via the North Pole from the Atlantic to the Pacific.

1960. Jacques Piccard and Don Walsh, in the *Trieste*, dive to the bottom of the Challenger Deep in the Marianas Trench, approximately 11 km (7 mi) below the surface.

1960. The nuclear-powered *USS Triton* becomes the first submarine to circum-navigate the globe.

1961. Harry Hess and Robert Dietz propose the theory of 'seafloor spreading'.

1962. Hans Keller and Peter Small descend to 330 m (1000 ft) in an open diving bell using a secret mixture of gases – the deepest depth by a diver without a pressure suit. Small dies during the attempt.

1977. Using the piloted Deep Submergence Vehicle (DSV) *Alvin*, an American team dives to a volcanic rift in the Pacific and discovers warm springs teeming with previously unknown species, including tube worms, snake-like creatures standing upright in long tubes.

1978. The first 'remote-sensing' oceanographic satellite (Seasat-A) is launched to study the ocean.

1979. An American team exploring the Gulf of California with *Alvin* finds mineral chimneys that discharge water hot enough to melt lead.

1980. Scientists propose that these seabed hot springs may be the birthplace of all life on Earth.

1984. American researchers diving off Florida in *Alvin* discover life swarming in cold springs, another new kind of deep-sea ecosystem.

1985. The remains of the *Titanic* are discovered by remotely operated cameras. The following year, *Alvin* makes 12 dives to photograph the wreck.

Right: The research submersible Alvin is used to observe and sample the deep-ocean bottom down to depths of 4 km (2.5 mi). The sampling arm (*bottom right*) is controlled by the pilot and samples are stored in the rack (*bottom left*). Cameras mounted on the hull take pictures of the seafloor.
Source: Gross, M. G. (1995).

1989. *Shinkai 6500*, a manned submersible, dives to 6527 m (21 414 ft) in the Japan Trench.

1992. Scientists propose that deep-sea ecosystems may harbor ten million species of life, far more than known on land.

1995. The Japanese-built *Kaiko* dives to the bottom of the Marianas Trench, recording a depth of 10 911 m (35 797 ft) and finds it alive with small animals.

1995. The US Navy releases seafloor gravity data, enabling civilian geographers to produce the first detailed public map of the ocean floor.

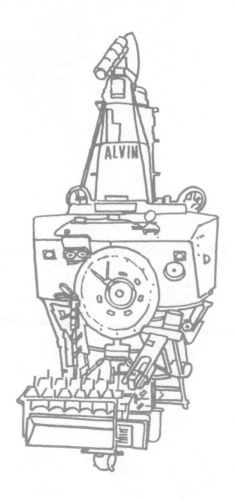

5

Ocean Trade

Ocean
transport
is
the most
inexpensive
way of
trading
bulk goods

As a result,
over 80%
of the
world's
trade
involves
transit via
the ocean

WORLD SHIPPING DISTANCES

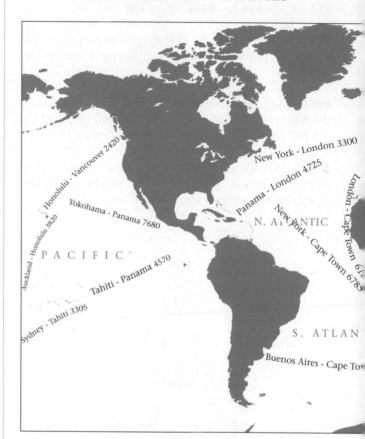

Honolulu - Vancouver 2420
Auckland - Honolulu 3820
Yokohama - Panama 7680
New York - London 3300
Panama - London 4725
New York - Cape Town 6785
London - Cape Town 61..
N. ATLANTIC
PACIFIC
Tahiti - Panama 4570
Sydney - Tahiti 3305
S. ATLAN
Buenos Aires - Cape To..

GROWTH OF MARITIME SHIPPING TRADE

	Maritime trade (in million metric tonnes)	Per head of world population (in metric tonnes)
1955	800	0.286
1975	3064	0.747
1995	4700	0.832
2000 (est.)	5690	0.940

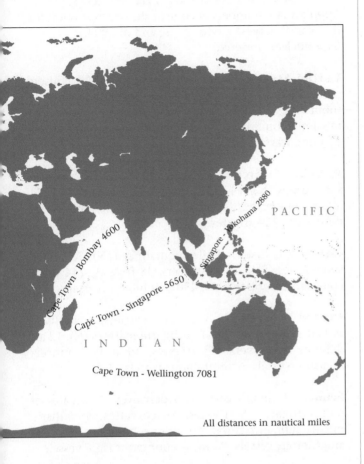

All distances in nautical miles

Source:
Couper, A., ed. (1983).

TEN LARGEST PORTS IN THE WORLD

Source:
Crook, G.
UNCTAD Secretariat
(1998).

Port	Goods (in million metric tonnes)	Latest year available
1. Singapore	306	(1995)
2. Rotterdam	294	(1995)
3. Chiba	193	(1992)
4. South Louisiana	172	(1996)
5. Houston	134	(1996)
6. Hong Kong	127	(1995)
7. Nagoya	124	(1995)
8. New York/New Jersey	119	(1996)
9. Antwerp	108	(1995)
10. Yokohama	108	(1995)

6

Geopolitics

The largest
navy in
the world
today is
that of the
United
States of
America

MAJOR NAVIES OF THE WORLD

Approximate numbers of combat ships in operation for the world's largest navies at the end of 1997: (N = nuclear-powered)

United States: 12 aircraft carriers (8N), 29 cruisers (2N), 56 destroyers, 42 frigates, 18 ballistic missile submarines (N), 61 attack submarines (N), 4 special mission submarines (N), 12 amphibious assault ships, 24 mine warfare ships.

Russia: 1 aircraft carrier, 39 ballistic missile submarines (N), 14 cruise missile submarines (N), 74 attack submarines (approx. 50N), 6 battle cruisers, 29 destroyers, 114 frigates.

China: 2 ballistic missile submarines (1N), 75 attack submarines (5N), 18 destroyers, 35 frigates, 144 mine countermeasures vessels, 430 amphibious vessels.

United Kingdom: 3 aircraft carriers, 3 ballistic missile submarines (N), 13 other submarines (N), 12 air defence destroyers, 23 frigates, 19 mine counter-measure vessels, 6 amphibious vessels.

France: 2 aircraft carriers, 5 destroyers, 9 frigates, 9 patrol frigates, 8 frigates, 16 corvettes, 4 ballistic missile submarines (N), 10 other submarines (6N), 10 amphibious vessels, 21 mine countermeasure vessels.

Japan: 13 destroyers, 44 frigates, 1 corvette, 15 submarines, 7 amphibious vessels, 35 mine counter-measure vessels.

Italy: 1 aircraft carrier, 1 helicopter cruiser, 4 destroyers, 12 frigates, 4 patrol frigates, 8 corvettes, 8 submarines, 3 amphibious vessels, 13 mine countermeasure vessels.

Germany: 3 air defence destroyers, 12 frigates, 14 submarines, 39 mine countermeasure vessels.

1. Bab el Mandab. Connecting the Red Sea with the Gulf of Aden and the Arabian Sea, this strait is a bottleneck for sea traffic between the Mediterranean Sea and the Indian Ocean. It is surrounded by Yemen, Djibouti and Eritrea.

2. Bosporus. Only 0.9 km (0.5 nm) wide at its narrowest point north of Istanbul, this 17-mile long strait connects the Black Sea with the Sea of Marmara. One of the most difficult to navigate waterways in the world; there were 155 total collisions in the Strait between 1988 and 1992. A major route for oil exports from Russia, the Caspian region, and Central Asia.

3. Dardanelles. Located in Turkey; separates the Gallipoli Peninsula and Asia Minor, and connects the Aegean Sea with the Sea of Marmara. Of great importance economically to Turkey, Russia and the Black Sea states.

4. Dover Strait. Connects the English Channel with the North Sea. The busiest of all straits used for international navigation.

5. Strait of Hormuz. Links the Arabian Gulf with the Gulf of Oman and the Arabian Sea; 14 million barrels per day of crude oil pass through it.

6. Strait of Malacca. Situated between Malaysia and Indonesia; connects the northern Indian Ocean with the South China Sea and Pacific Ocean. One-third of the world's ships pass through the Strait of Malacca and the nearby Strait of Singapore, the major route for ships travelling from the Arabian Gulf to Japan.

7. Strait of Gibraltar. Separates Africa from Europe. The only natural access by sea into the Mediterranean.

There were 155 total collisions in the Bosporus between 1988 and 1992

8. Suez Canal. 160 km (100 mi) long, the Suez Canal was originally opened on 17 November 1869. A long-term programme for deepening the canal began in the 1950s; by 1999, the maximum permissible draught is expected to be 19 m (62 ft), and by 2010, 22 m (72 ft).

The canal has twice been the flashpoint for military confrontation: in 1956, when Egypt nationalized the canal, planning to use tolls to pay for construction of the Aswan High Dam, Israel, France and Britain launched an abortive military campaign to re-establish international control of the area. Fifteen ships in the canal were sunk during the Arab–Israeli war in June 1967; the canal was closed for repairs until 1975.

The volume of trade passing through the canal constitutes 7% of the world trade volume.

9. Panama Canal. Opened on 15 August 1914, the Panama Canal is 80.5 km (50 mi) long. Its construction cut the distance by sea between the east and west coasts of the USA by 8000 mi; it is also a vital link from Europe to the west coast of the Americas and the Pacific.

Approximately 14 000 ships use the canal each year, with an average transit time of 8 to 10 hours; however, because of increasing congestion, ships can often wait up to 18 hours before transiting the canal. Various options for expansion, including the construction of a new sea-level canal across Panama, are being considered.

7% of the world trade volume passes through the Suez Canal

7

Ocean Resources

Over the past 50 years, water use world-wide has increased four-fold

Over the past 50 years, water use world-wide has increased four-fold. However, greatly increased levels of water extraction, poor water management, and the contamination of groundwater supplies as a result of pollution from, for example, intensive livestock production and industrial fertilizer and pesticide use, are putting increasing pressure on fresh-water supplies.

Such problems will probably be further exacerbated as a result of global climate change. Some parts of the world will not only receive less rainfall but the increase in temperature will mean that a higher proportion of the water falling on the earth's surface will evaporate. In addition, rising sea-levels are likely to cause further saltwater intrusion into groundwater supplies.

Accordingly, this has prompted a greatly increasing interest in the ocean as a source of fresh water.

World-wide, there are now approximately 10 300 water desalination plants (with a capacity of at least 100 cu. m/day (3530 cu. ft/day), with a total production capacity in 1995 of 19 200 000 cu. m (678 450 000 cu. ft) of fresh water daily.

This is a significant increase over the past 25 years: in 1970, total production was just 1 330 000 cu. m/day (47 000 000 cu. ft/day).

Small-scale desalination units are common on board of ocean-going vessels and in some small coastal and island communities, for example, in the Aegean and Caribbean. However, on a larger scale to meet the demands of large urban and industrial centres, for example, or to supply water for irrigation, central desalination plants require such large amounts of energy that they are, at present, rarely cost effective.

MINERALS AND MINERAL EXPLOITATION
(Excluding off-shore oil and gas)

Ocean waters are potentially the most significant source of minerals for sustainable recovery on earth. Indeed, the ocean is estimated to contain 80% of the global resources potential.

However, at present, the only elements commercially extracted from sea water on a large scale are magnesium and magnesium compounds (used in the chemical industry), bromine (primarily used in petroleum products), salt and heavy water (deuterium oxide).

Continental shelves contain sands and gravels, shell and coral sands, used primarily in the construction industry, and quantitatively the most important marine mineral deposits mined. Other continental shelf mineral resources include:

● mineral sands (containing gold, platinum, diamonds and other gem stones, tin, titanium, iron, zirconium, chromium, and rare earth oxides);

● phosphorite (containing dolomite, silica, magnesium-rich clays, glauconite, manganese, calcite, and organic matter), used for the manufacture of phosphate fertilizers and certain phosphorous-based chemicals;

● hard rock deposits (containing coal, phosphates, carbonates, potash, ironstone, limestone, metal sulfides, and metal salts);

Ocean basin and seafloor deposits include:
● nodules (containing manganese, cobalt, nickel, copper, etc.)
● crusts (containing phosphorite, cobalt, and manganese)
● mounds and stacks (containing metal sulphides of copper, gold, lead, silver, etc.);

Major marine mineral industries/locations:
● sand and gravel mining – Japan, North Sea
● diamond mining – South Africa, Namibia
● salt production – various countries
● magnesium compounds – USA
● magnesium metal – USA.

Source:
Food and Agriculture
Organization of the United
Nations (FAO), Rome.

UTILIZATION OF WORLD FISH PRODUCTION DURING 1986–1995
Disposition ('000 tonnes live weight)

	1986	1987	1988	1989	1990	1991	1992	1993	1994	1995
Total world catch	93 051	94 687	99 336	100 660	97 972	97 797	100 177	103 172	110 538	112 910
For human consumption	64 090	66 717	69 824	70 407	70 080	69 123	71 968	73 937	77 072	82 126
Marketing fresh	20 023	22 117	23 761	22 621	21 307	20 398	24 276	24 819	26 142	29 857
Freezing	22 419	22 646	23 711	24 517	24 908	24 661	25 274	26 138	27 363	28 095
Curing	10 036	10 188	10 206	10 618	10 944	10 950	9 913	10 154	10 816	11 361
Canning	11 612	11 766	12 146	12 651	12 921	13 114	12 505	12 826	12 751	12 813
For other purposes	28 961	27 970	29 512	30 253	27 892	28 674	28 209	29 235	33 466	30 784
Reduction	27 361	26 420	27 862	28 653	26 392	27 074	26 409	27 435	31 666	28 984
Miscellaneous	1 600	1 550	1 650	1 600	1 500	1 600	1 800	1 800	1 800	1 800

MARINE POLLUTION

8
Threats to the Oceans

Source:
Weber, P. (1993).

Type	Causes/Sources
Nutrients	Run-off from sewage, farming, forestry; airborne emissions from power plants, cars, etc.
Sediments	Erosion from mining, forestry, farming, coastal development, etc.
Persistent toxins	Industrial discharge, wastewater, pesticides, etc. (PCBs, heavy metals, etc.)
Oil	Cars, industry, tankers and shipping, off-shore oil drilling
Plastics	Fishing nets; cargo and cruise ships; litter; wastes from industry and landfills
Radioactive isotopes	Discarded nuclear submarines and military wastes; power plant emissions, industrial waste

CHEMICAL POLLUTANTS

• Number of new industrial chemicals submitted yearly for evaluation under the US Toxic Substances Control Act: 1500–2000.

• Number of chemicals listed in the *European Inventory of Existing Chemical Substances*: > 100 000.

• Amount of nitrogen emitted to the atmosphere each year from fossil fuel combustion: approx. 20 million tonnes; projected estimate for 2020: approx. 46 million tonnes.

• Amount of nitrogen mobilized each year from biomass burning, land-clearing and wetland drainage: approx. 70 million tonnes.

• Amount of nitrogen produced industrially each year for fertilizer: approx. 80 million tonnes; projected estimate for 2020: 134 million tonnes.

• Amount of phosphorus used annually as fertilizer: 30 million tonnes.

• Amount of municipal effluent and industrial waste-water and spent cooling water discharged into US coastal waters: 80 000 million litres (17 600 million gallons) a day.

• Estimated amount of petroleum hydrocarbons released into the ocean each year:
   (a) by natural processes (marine seeps, sediment erosion): 250 000 tonnes.
   (b) by human activities (e.g. spills, runoff, discharge): 2 500 000 tonnes.

The burning of fossil fuels emits approx. 20 million tonnes of nitrogen into the air each year

DAMS, RESERVOIRS AND FRESHWATER FLOW

● Maximum water surface areas of dam reservoirs: approximately 500 000 sq. km (195 000 sq. mi).

● Number of large dams globally: ~ 39 000

● Estimated percentage of global river flow to the sea that has been dammed or diverted:
 – in the early 1990s: 13%
 – projected for early Twenty-first century: 20%

● Percentage of Nile River flow reaching the Mediterranean after construction of the High Aswan Dam: <3%

● Percentage reduction in sediment loading to the Indus Delta (Arabian Sea, Pakistan) after construction of barrages: >80%

Examples of major dam impacts on coastal environments:

● Reduction of sediment outflow from the Danube River has resulted in the erosion of the Danube Delta (Black Sea) by up to 17 m/yr (56 ft/yr).

● The near-complete cessation of sediment flow from the Nile has resulted in the erosion of the Nile Delta (Mediterranean) with over 100 m (330 ft) at certain places in certain years.

● Reduction of nutrient inputs into the Mediterranean from the Nile River resulted in a 75% reduction in Egypt's annual shrimp landings and a 90% reduction in its sardine landings.

● Reductions in riverine silica loading into the Black Sea, thus causing altered nutrient ratios, have probably prompted the emergence of novel phytoplankton blooms (or 'red tides').

● One of the main contributors to the extinction of around 250 anadromous salmon and steelhead and sea-run cutthroat trout populations along the US and Canadian Pacific coast.

There are 39 000 large dams globally

197

INTRODUCED SPECIES

● Current primary sources of non-indigenous species introduction into estuarine and coastal waters:
  - ballast water discharge
  - aquaculture and aquarium-related introductions

● Past primary sources of non-indigenous species introductions into estuarine and coastal waters:
  - transoceanic canals (i.e. Suez Canal, Panama Canal)
  - attachment to ships, hulls

● Number of species of marine and estuarine organisms identified in the ballast water of ships in port: > 350

● Estimated number of species in motion at any one time in ballast water of ships on the oceans: >3000

● Number of toxic phytoplankton cysts discovered in the ballast water of one ship in an Australian port: ~300 million

● Estimated losses to fisheries in the Black Sea as a result of the introduced comb jelly *Mnemiopsis leidyi*: US$ 300 million

● Number of known non-indigenous species in the Central and Northern Baltic Sea: 35

● Number of known non-indigenous plants and animals in San Francisco Bay: >150

● Densities of the introduced Chinese clam in various bottom areas of San Francisco Bay: 10 000 per sq. m (900 per sq. ft).

9

Marine Protected Areas (MPAs)

• Number of MPAs globally: >1300

• Percentage of the planet's land surface designated as protected: > 6%

• Percentage of the planet's marine area designated as protected: <1%

• Year in which the concept of marine parks was first publicized at the international level: 1962 (at the First World Conference on National Parks)

The Great Barrier Reef
(area marked above) is the largest
living feature on earth and is
clearly visible from the moon.

Less than 1% of the planet's marine area is designated as protected

• Largest MPA: ~ 350 000 sq. km (136 500 sq. mi), (Great Barrier Reef Marine Park (GBRMP), Australia)

• Percentage of the Great Barrier Reef incorporated in the GBRMP: 99%

• Number of individual reefs in the GBRMP: ~ 2500

• Number of species recorded in the GBRMP:

(a) hard coral	> 300
(b) fishes	~1500
(c) birds	~ 240
(d) molluscs	>4000
(e) sea turtle	6

POTENTIAL BENEFITS OF MPAs

- protection of marine species at certain stages of their life cycle

- protection of fixed, critical, habitats (e.g. coral reefs, estuaries)

- protection of cultural and archeological sites

- protection of local and traditional sustainable marine-based lifestyles and communities

- provision of space to allow shifts in species distributions in response to climate and other environmental changes

- provision of a refuge for recruits to commercial fisheries

- provides a framework for resolving multiple stake-holder conflicts

- provides models for integrated coastal zone management

- provision of revenue and employment

- provision of areas for scientific research, education, and recreation

EXISTING AND PROPOSED MARINE
PROTECTED AREAS (MPA)
(All numbers approximate and subject to change)

Marine region	Existing	Proposed
Antarctic	17	0
Arctic	16	29
Mediterranean	53	57
North West Atlantic	89	12
North East Atlantic	41	12
Baltic	43	22
Wider Caribbean	96	11
West Africa	42	59
South Atlantic	19	4
Central Indian Ocean	15	22
Arabian Seas	19	15
East Africa	50	24
East Asian Seas	92	44
South Pacific	65	58
North East Pacific	167	1
North West Pacific	190	1
South East Pacific	19	1
Australia and New Zealand	291	20
Total	1324	392

Source:
Kelleher, G., Bleakley, C.
and Wells, S. (1995).

Annex A

Sources

1. Basic Facts:

(1) Broecker, W.S. 1991. The great ocean conveyor. *Oceanography*, 4: 79-89. (2) Carson, R. 1951. *The Sea Around Us*. Oxford University Press, New York. (3) Couper, A., ed. 1983. *The Times Atlas of the Ocean*. Van Nostrand Reinhold, New York. (4) Garrison, T. 1996. *Oceanography: An Introduction to Marine Science*. International Thomson Publishing, London. (5) Gross, M.G. and Gross, E. 1995. *Oceanography: A View of Earth*. Prentice-Hall, New York. (6) Groves, D.G. and Hunt, L.M. 1980. *Ocean World Encyclopedia*. McGraw-Hill, New York. (7) Harvey, J.G. 1979. *Atmosphere and Ocean: Our Fluid Environments*. The Artemis Press, Sussex, UK. (8) Thurman, H. 1997. *Introductory Oceanography*. Simon & Schuster, New York. (9) World Meteorological Organization. 1988. Report WMO/TD No 215.

Information also from: Smithsonian Institution; United States Naval Meteorology and Oceanography Command; Woods Hole Oceanographic Institution.

2. Life in the Oceans:

(1) Broad, W.J. 1997. *The Universe Below*. Simon & Schuster, New York. (2) Briggs, J.C. 1994. Species diversity: land and sea compared. *Systematic Biology*, 43: 130–135. (3) Cone, J. 1991. *Fire Under the Sea*. Quill, New York. (4) Couper, A., ed. 1983. *The Times Atlas of the Oceans*. Van Nostrand Reinhold, New York. (5) Grassle, J.F. and Maciolek, N.J. 1992. Deep-sea species richness: regional and local diversity estimates from quantitative bottom samples. *American Naturalist*, 139: 313–341. (6) Gray, J.S. et al. 1997. Coastal and deep-sea benthic diversities compared. Marine *Ecology - Progress Series*, [159]: 97–103. (7) Gross, M.G. and Gross, E. 1995. *Oceanography: A View of Earth*. Prentice-Hall, New York. (8) Koslow, J.A., Paxton, J.R. and Williams, A. 1997. How many demersal fish species in the deep-sea? A test of a method to extrapolate from local to global diversity. *Biodiversity and Conservation*, 6 [11]: 1523–1532. (9) Lalli, C.M. and Parsons, T.R. 1993. Biological *Oceanography: An Introduction*. Pergamon Press, Oxford. (10) Marshall, N.B. 1980. *Deep-sea Biology*. Garland STPM Press, New York. (11) May, R.M. 1994. Biological diversity: differences between land and sea. *Philosophical Transactions of the Royal Society of London*, Part B, 343: 105–111. (12) Pace, N.R. 1997. A molecular view of microbial diversity in the biosphere. *Science*, 276: 734–740. (13) Poore, G.C.B. and Wilson, G.D.F. 1993. Marine species richness. *Nature*, 361: 597–598. (14) Reaka-Kudla, M.L. 1997. The global biodiversity of coral reefs: a comparison with rain forests. In Reaka-Kudla, M.L., Wilson, D.E. and Wilson E.O. (eds). *Biodiversity II: Understanding and Protecting our Biological Resources*. Joseph Henry Press, Washington, D.C. (15) Sournia, A. et al. 1991. Marine phytoplankton: how many species in the world ocean? *Journal of Plankton Research*, 13 [5]: 1093–1099. (16) United Nations Environment Programme. 1995. *Global Biodiversity Assessment*. Cambridge University Press, Cambridge. (17) Van Dover, C.L. 1997. *Deep Ocean Journeys*. Addison-Wesley, New York. (18) Waller, G., ed. 1996. *SeaLife: A Complete Guide to the Marine Environment*. Smithsonian Institution, Washington, D.C.

3. Coasts, Islands and Populations:

(1) Anderson, A.B. et al. 1978. The decline of macrofauna in the deeper parts of the Baltic proper and the Gulf of Finland. *Kieler Meeresforsch*, 4: 23–52. (2) Bird, E.C.F. 1985. *Coastline Changes: A Global Review*. Wiley-Interscience, Chichester. (3) Bird, E.C.F. *Submerging Coasts: The Effects of a Rising Sea-Level on Coastal Environments*. Wiley, Chichester. (4) Cohen, J.E. et al. 1997. Estimates of coastal populations. *Science*. 278 [5341]: 1211–1212. (5) Houghton, J. 1997. *Global Warming: The Complete Briefing*. Cambridge University Press, Cambridge. (6) IUCN 1993. *Reefs at Risk: A Program for Action*. IUCN, Gland. (7) Mathieson, A.C. and Nienhuis, P.H. 1991. *Intertidal and Littoral Ecosystems*. Elsevier, Amsterdam. (8) *National Geographic Atlas of the World*. 1996. National Geographic Society, Washington, D.C. (9) Pernetta, J.C. and Elder, D.L. 1992. Climate, Sea-level rise and the coastal zone: management and planning for coastal changes. *Ocean and Coastal Management*, 18: 113–160. (10) Price, J.H., Guthrie, D.E.G. and Farnham, W.F. 1980. *The Shore Environment. Vol. 2: Ecosystems*. Academic Press, New York. (11) Short, F.T. and Wyllie-Echeverria, S. 1996. Natural and human-induced disturbance of sea grasses. *Environmental Conservation*. 23 [1]: 17–27. (12) Turner, R.E. and Rabelais, N.N. 1994. Coastal eutrophication near the Mississippi River Delta. *Nature*, 14 April. (13) Turner, R.K., Subak, S. and Adger, W.N. 1996. Pressures, trends and impacts in coastal zones: interactions between socioeconomic and natural systems. *Environmental Management*. 20 [2]: 159–173. (14) Waller, E., ed. 1996. *SeaLife: A Complete Guide to the Marine Environment*. Smithsonian Institution, Washington, D.C. (15) Watson, R.T., Zinyowera, M.C. and Moss, R.H., eds. 1996. *Climate Change 1995*. Impacts, adaptations and mitigation of climate change: Scientific-technical analyses. Contribution of Working Group II to the Second Assessment Report of the Intergovernmental Panel on Climate Change. Cambridge University Press, Cambridge. (16) Watzman, H. 1995. Building boom destroys Israel's delicate dunes. *New Scientist*, 10 June, 8.

4. Ocean Exploration and Discovery:

(1) Ballard, R. and McConnell, M. 1995. *Explorations: My Quest for Adventure and Discovery Under the Sea*. Hyperion, New York. (2) Beaglehole, J.C. 1974. *The Life of Captain James Cook*. A & C Black, London. (3) Beebe, W. 1934. *Half Mile Down*. Harcourt, Brace, New York. (4) Boorstin, D. 1983. *The Discoverers: A History of Man's Search to know his World and Himself*. Random House, New York. (5) Borgese, E.M., ed. 1992. *Ocean Frontiers: Explorations by Oceanographers on Five Continents*. Harry N. Abrams, New York. (6) Broad, W.J. 1997. *The Universe Below: Discovering the Secrets of the Deep Sea*. Simon & Schuster, New York. (7) Cone, J. 1991. *Fire Under the Sea: The Discovery of the Most Extraordinary Environment on Earth, Volcanic Hot Springs on the Ocean Floor*. Quill, New York. (8) Cousteau, J. 1979. *The Ocean World*. Harry N. Abrams, New York. (9) Delpar, H. 1979. *The Discoverers: An Encyclopaedia of Explorers and Exploration*. McGraw-Hill, New York. (10) Earle, S. 1995. *Sea Change: A Message of the Oceans*. Putnam, New York. (11) Earle, S. and Giddings, A. 1980. Exploring the Deep Frontier: The Adventure of Man in the Sea. *National*

Geographic Society, Washington, D.C. (12) Gross, M.G. and Gross, E. 1995. *Oceanography: A View of Earth*. Prentice-Hall, New York. (13) Hendrickson, R. 1983. *The Ocean Almanac*. Doubleday, New York. (14) Heyerdahl, T. 1950. *The Kon-Tiki Expedition*. Allen & Unwin, London. (15) Royal Geographical Society, 1997. *The Oxford Atlas of Exploration*. Oxford University Press, New York. (16) Mill, H.R. 1905. *The Siege of the South Pole*. Alston Rivers, London. (17) Severin, T. et al. 1995. *The China Voyage: Across the Pacific by Bamboo Raft*. Addison-Wesley, New York. (18) Waller, G., ed. 1996. *SeaLife: A Complete Guide to the Marine Environment*. Smithsonian Institution, Washington, D.C. (19) Whitfield, P. 1996. *The Charting of the Oceans: Ten Centuries of Maritime Maps*. The British Library, London.

5. Ocean Trade:

(1) Couper, A., ed. 1983. *The Times Atlas of the Oceans*. Van Nostrand Reinhold, New York. Additional information from Gary Crook of the UNCTAD Secretariat.

6. Geopolitics:

(1) Couper, A., ed. 1983. *The Times Atlas of the Oceans*. Van Nostrand Reinhold, New York. (2) US Naval and Shipbuilding Museum. *The World's Navies: The Naval Institute's Guide to Combat Fleets of the World*. (www.uss-salemorg/worldnav). Naval Institute Press, Annapolis.

Additional information from: Royal Navy; United States Navy; Interoceanic Region Authority, Panama; Egyptian State Information Service; St. Lawrence Seaway Authority; Strategic Studies Institute, US Army War College; United States Energy Information Administration.

7. Ocean Resources:

(1) Couper, A., ed. 1983. *The Times Atlas of the Oceans*. Van Nostrand Reinhold, New York. (2) Cruickshank, M.J. 1992. Marine mineral resources. In *Encyclopedia of Earth System Science. Vol. III*. Academic Press. (3) Houghton, J.G. 1997. *Global Warming: The Complete Briefing*. Cambridge University Press, Cambridge. (4) Lanz, K. 1995. *The Greenpeace Book of Water*. David & Charles, Newton Abbot, UK. (5) Tolba, M.K. and El Kholy, O.A., eds. 1992. *The World Environment 1972-1992*. Chapman & Hall, London. (6) Wangnick Consulting GMBH, ed. 1996. *IDA Worldwide Desalting Plants Inventory*. Report No. 14. Gnarrenburg. (7) World Resources Institute, 1992. *World Resources 1992-1993*. Oxford University Press, New York.

Additional information from: Cruickshank, M.J. (University of Hawaii) [personal communication]; Food and Agriculture Organisation of the United Nations (FAO), Rome.

8. Threats to the Oceans:

(1) Carlton, J.T. 1985. Transoceanic and interoceanic dispersal of coastal marine organisms: the biology of ballast water. *Oceanography and Marine Biology Annual Review*, 23: 313–371. (2) Carlton, J.T. and Geller, J.B. 1993. Ecological roulette: the global transport of non-indigenous marine organisms. *Science*, 236: 78–82. (3) Carlton, J.T. et al. 1990. Remarkable invasion of San Francisco Bay (California, USA) by the Asian clam *Potamocorbula amurensis*.

I. Introduction and dispersal. *Marine Ecology Progress Series*, 66: 81–94. (4) Commission of the European Communities; FAO. 1993. *Food and Agriculture Yearbook. 1992.* Statistical Series 112, FAO, Rome. (5) Dowidar, N.M. 1988. Productivity of the South-eastern Mediterranean. In *Natural and Man-Made Hazards*, El-Sabh, M.I. and Murty, T.S., eds. Reibel, The Netherlands. (6) Galloway, J.N. et al. 1995. Nitrogen fixation: atmospheric enhancement, environmental response. *Global Biogeochemical Cycles*, 9: 235–252. (7) Hallegraeff, G.M. and Bolch, C.J. 1991. Transport of toxic dinoflagellates via ships' ballast water. *Marine Pollution Bulletin*, 22: 27–30. (8) Howarth, R.W. and Marino, R. 1991. *Oil in the Oceans*. Greenpeace, Amsterdam. (9) Humborg, C. et al. 1997. Effect of Danube River dam on Black Sea biogeochemistry and ecosystem structure. *Nature*, 386: 385–388. (10) *International Water Power and Dam Construction Handbook.* 1994. Wilmington Business Publishing Co., Wilmington, UK. (11) Kelly, C.A. et al. 1994. Turning attention to reservoir surfaces, a neglected area in greenhouse studies. EOS, *Transactions of the American Geophysics Union*, 75: 332–333. (12) Lobert, J.M. et al. 1990. Importance of biomass burning in the atmospheric budget of nitrogen - containing trace gases. *Nature*, 34: 552–554. (13) Milliman, J.D. 1991. Flux and fate of fluvial sediment and water in coastal seas. In *Ocean Margin Processes in Global Change*. Mantoura, R.F.C., Martin, J–M. and Wollast, R., eds., John Wiley, Chichester. (14) Milliman, J.D., Quraishee, G.S., and Beg, M.A.A. 1984. Sediment discharge from the Indus River into the ocean: past, present and future. In *Marine Geology and Oceanography of the Arabian Sea*, Haq, B.U. and Milliman, J.D., eds. Van Nostrand Reinhold, New York. (15) National Oceanographic and Atmospheric Administration, 1993. Nonindigenous Estuarine and Marine Organisms, Proceedings of the Conference and Workshop, Seattle, Washington. (16) National Research Council, 1993. *Managing Waste-Water in Coastal Urban Areas*. National Academy Press, Washington, D.C. (17) National Research Council, 1995. *Oil in the Sea: Inputs, Fates and Effects*. National Academy Press, Washington, D.C. (18) Nehlsen, W., Williams, J.E. and Lichatowich, J.A. 1991. Pacific salmon at the crossroads: Stocks at risk from California, Oregon, Idaho and Washington. *Fisheries*, 16 [2]: 4–21. (19) Pearce, F. 1991. A dammed fine mess. *New Scientist*, 130 [1767]: 36-39. (20) Pringle, C. et al. 1993. Environmental problems of the Danube Delta. *American Scientist*, 81: 350–361. (21) Slaney, T.L. et al. 1996. Status of anadromous salmon and trout in British Columbia and Yukon. *Fisheries*, 21 [10]: 20–35. (22) Smith, S.E. and Abdel-Kadar, A. 1988. Coastal erosion along the Egyptian Delta. *Journal of Coastal Research*, 4: 245–255. (23) Wadie, W.F. and Abdel Razek, F.A. 1985. The effect of damming on the shrimp population in the south-eastern part of the Mediterranean Sea. *Fisheries Research*, 3: 323–335. (24) Weber, P. 1993. *Abandoned Seas: Reversing the Decline of the Oceans*. Worldwatch Institute, Washington, D.C. (25) World Resources Institute, 1992. *World Resources 1992–1993*. Oxford University Press, New York. (26) Zeeman, M. et al. 1995. US EPA regulatory practices on the use of QSAR for new and existing chemical evaluations. SAR QSAR *Environmental Research*, 3: 179–201.

9. Marine Protected Areas (MPAs):

(1) Agardy, M.T. 1994. Advances in marine conservation: the role of marine protected areas. *Trends in Ecology and Evolution*, 9 [7]: 267–270. (2) Bohnsack, J.A. and Ault, J.S. 1996. Management strategies to conserve marine biodiversity. *Oceanography*, 9 [1]: 73–82. (3) Eichbaum, W., Crosby, M., Agardy, M.T. and Laskin, S. 1996. The role of marine and coastal protected areas in the conservation and sustainable use of biological diversity. *Oceanography*, 9 [1]: 60–70. (4) Kelleher, G. 1985. The Great Barrier Reef Marine Park. In *Marine Parks and Conservation: Challenge and Promise*. Vol. 2., Lien, J. and Graham, R., eds. National and Provincial Parks Association, Toronto. (5) Kelleher, G., Bleakley, C. and Wells, S. 1995. *Priority areas for a global representative system of marine protected areas*. Four-volume report to the World Bank Environment Department, Washington, D.C. (6) Ray, G.C. 1962. Inshore marine conservation. In *First World Congress on National Parks*, Adams, A.B., ed. National Park Service, Washington, D.C. (7) Ticco, P. 1995. The use of marine areas to preserve and enhance marine biological diversity: a case study approach. *Coastal Management*, 23: 309–314. (8) World Monitoring Center, 1992. *Global Biodiversity: Status of the Earth's Living Resources*. Chapman & Hall, London.

REGIONAL OR NATIONAL INPUTS

In preparing its Final Report, the IWCO drew upon the substance of regional or national inputs, through meetings and otherwise, from various places in the world. The most relevant inputs were, in chronological order, the following:

● *Summary Report, Regional Hearing for the World Commission on the Seas & Oceans*, IOI China Operational Center, Tianjin, China, 9–11 May 1995.

This hearing, at the IOI China Operational Center, was strongly supported by the Chinese Government. It was attended by 55 participants including experts in international law and law of the sea, a captain of an ocean-going ship, scholars in marine science and technology, marine environmental protection, marine and coastal zone management, marine resources development, marine mapping, marine fisheries, desalination and utilization of sea water, navigation, harbor and wharf management, marine data processing.

● *Summary Report, International Hearing on Ocean Affairs*, Japanese Commission on the Oceans, Yokohama, Japan, 11 September 1995.

The International Hearing on Ocean Affairs, held on 11 September 1995 in Yokohama, was sponsored by the Japanese Commission on the Oceans and supported by the Nippon Foundation and Yokohama City. More than 300 participants attended the hearing, which was open to the public. It was followed by a day of field study: a cruise planned and sponsored by the Port Authority with a lecture on the Port, its history since the mid-nineteenth century, present condition and blue-print for the twenty-first century.

● *Report, Hearings for Oceania*, IOI–South Pacific and the University of the South Pacific, Suva, Fiji, September–October 1995.

A series of eight hearings were held in various countries of the Pacific Region – in Fiji, Western Samoa, New Zealand, Australia, Solomon Islands, Tuvalu, Marshall Islands and Kiribati – between 28 September and 31 October 1995. The hearings were attended by approximately 200 people from all walks of life. These included fishers, public servants, politicians, scientists, students,

Annex

B

Regional
or
National
Inputs

representatives of non-governmental organizations (NGOs) and interested members of the public.

● *Contribution of the Use of the Oceans and their Resources*, Independent National Commission on the Oceans, Rio de Janeiro, Brazil,19 April 1996.

The Independent National Commission on the Oceans (INCO) – Brazil was established on 19 April 1996. Presided by the Minister of Science and Technology, Dr José Israel Vargas, who is also a Vice-President of the IWCO, the INCO is constituted of 18 highly representative members, including three former Ministers of State, scientists, legal counsellors, and representatives of sea transport and ship-building, fishery, exploration of marine mineral resources and environment. Its first publication contains articles on maritime transportation, fisheries, off-shore oil exploitation, exploitation of minerals, marine pollution, and coastal ecosystems.

● Contribution du Sénégal à la CIMO par Monsieur Alassane Dialy Ndiaye, Ministre de la Pêche et des Transports Maritimes du Sénégal, Mai 1996.

This contribution aims at showing, in various sectors, the importance of the oceans and of the adjacent coastal zone for the life of the Senegalese. The accent is put on the principal resources exploited, the systems of exploitation, the state of the ecosystems and of the principal resources concerned, and on the structures put in place to assure implementation of the best strategies of exploitation. The efforts needed to take the marine environment into account in the framework of Agenda 21 are also underlined.

● Proceedings, National Seminar on *Coastal Ocean Regime and Developing Countries*, Department of Ocean Development, New Delhi, 21 June 1996.

With the entering into force of the United Nations Convention on the Law of the Sea, it was felt necessary to understand some of its implications. To this end, a National Seminar was organized by the Department of Ocean Development, Government of India. The Seminar focused on five specific issues: (i) The need for awareness about the oceans; (ii) The legal regime for coastal states and islands; (iii) Ocean technology and its relevance for the developing countries; (iv) Coastal marine protection and management; and (v) Deep–sea fishing and its impact on coastal fisheries. It had 65 participants from government, universities and NGOs.

• *Report to the Independent World Commission on the Oceans, Voice for the Oceans,* International Ocean Institute (India), Madras, October 1996.

IOI India conducted a 'Samudra Manthan' with a group of people in India and the Indian Ocean Region, inviting them to think about the oceans and come out with their views, concerns, suggestions. This was done through hearings in India and a survey in the Region. The number reached was only a few hundred, but they were all people connected with, and concerned about, the oceans – through work, environmental activism or sheer love for the sea.

• *Final Report, Canadian Ocean Assessment: A Review of Canadian Ocean Policy and Practice,* International Ocean Institute, Halifax, October 1996.

The project was essentially one of information-gathering, involving the solicitation of public opinion from academia, from government departments, and to some extent from the grass-roots level, to provide a current status review of the oceans and of oceans management policy and practices. Four information-gathering components were used in the project, including three public hearings (with, in total, 76 different participants), individual surveys, submitted briefs, and current organization reports.

• *Opinions and Attitudes of Brazilians towards the Sea,* for the Independent National Commission on the Oceans – Brazil, by GALLUP, August 1997.

During June and July 1997, the 'Instituto GALLUP de Opinião Pública' carried out an exclusive research work for the Independent National Commission on the Oceans (INCO) – Brazil, with the aim of collecting data about the opinions and attitudes of the Brazilians towards the sea. The objective was to provide data on: (i) Importance of the sea for the Brazilians; (ii) Political and economic aspects of the sea; and (iii) Merchant marine, navy, naval industry and ports. The research was carried out nationally in urban areas of Brazil.

• *Uses of the Oceans in the XXI Century – A Brazilian Approach*, Report of the INCO – Brazil, January 1998.

This report, in following the format of the Commission's Final Report, concerns:

(i)   Peaceful uses of the oceans, sovereignty and security;
(ii) Sustainable uses of the oceans;
(iii) Possibilities and challenges of science and
      technology; and
(iv) Awareness, partnership and solidarity on uses of
      the oceans.

In each case, it focuses also on the Brazilian situation.

• *Rapport Final, Rencontre Internationale de Dakar*, 25–27 March 1998.

This meeting was organized in the context of the work of IWCO and included, in addition to Senegalese, invited experts from Cape-Verde, Morocco, Gambia, and Guinea, and from international organizations. Two major themes were developed:

(i) African and major ocean issues, traditions, myths and realities;
(ii) Artisanal fisheries in Africa, economic and socio-cultural aspects.

At the end of the meeting, a Final Resolution was adopted, presenting suggestions relating to an African vision of the ocean and its actualities.

As shown by these brief background descriptions, the purpose and scope of these documents is far from uniform. Nevertheless, taken as a whole they do provide a fairly detailed panoply of perceptions and experiences at the national and regional levels which have served to enrich the content of the Final Report of the Commission. Common to most of these documents is an affirmation of the need to: ratify and/or implement the Law of the Sea Convention; seek cooperation to help overcome the lack of knowledge about the ocean and the resources available in the EEZs in question; and adopt an integrated approach to the development and protection of coastal areas.

SELECTED SOURCES

General

Anand, R.P. 1983. *Origin and Development of the Law of the Sea: History of International Law Revisited*. Martinus Nijhoff, The Hague.

Broad, W.J. 1997. *The Universe Below: Discovering the Secrets of the Deep Sea*. Simon & Schuster, New York.

Commission on Global Governance. 1995. *Our Global Neighbourhood*. Oxford University Press, Oxford.

Couper, A.D., ed. 1983. *The Times Atlas of the Oceans*. Times Books, London.

Dupuy, R-J. and Vignes, D., ed. 1985. *Traité du Nouveau Droit de la Mer*. Bruylant, Brussels.

GESAMP, 1990. *The State of the Marine Environment*. Blackwell, Oxford and London.

Green Globe Yearbook of International Cooperation on Environment and Development, 1994. Fridtjof Nansen Institute, ed. Oxford University Press, New York.

International Ocean Institute, 1978–1996. *Ocean Yearbook*, Mann Borgese E., Ginsburg N., and Morgan J.R., eds., University of Chicago Press, Chicago, vols. 1–12.

Juda, L. 1996. *International Law and Ocean Use Management: The Evolution of Ocean Governance*. Routledge, London.

Kimball, L.A., Johnston, D.M., Saunders P.M. and Payoyo P., eds. 1995. *The Law of the Sea: Priorities and Responsibilities in Implementing the Convention*. IUCN, Gland, Switzerland.

Lévy, J-P. 1995. Les Nations Unies et la Convention de 1982 sur le Droit de la Mer, *Revue Belge de Droit International*, Editions Bruyland, No. 1.

De Marco, G. and Bartolo, M. 1997. *A Second Generation United Nations for Peace in Freedom in the 21st Century*. Kegan Paul International, London and New York.

Annex C

Annex C

Selected Sources

Papon, P. 1996. *Le Sixième Continent: Géopolitique des Océans*. Editions Odile Jacob, Paris.

Pernetta, J. 1994. *Philip's Atlas of the Oceans*. Reed International Books Ltd., London.

United Nations. 1993. *Agenda 21: Programme of Action for Sustainable Development, Rio Declaration on Environment and Development, Statement of Forest Principles*. UNCED, 3–14 June 1992. UN Publication, Sales No. E.93.I.11, New York.

Smithsonian Institution. 1996. *Sea Life: A Complete Guide to the Marine Environment*. Smithsonian Institution Press, Washington, DC.

Weber, P. 1993. *Abandoned Seas: Reversing the Decline of the Oceans*, Paper 116, Worldwatch Institute, Washington, DC.

World Commission on Environment and Development (The Brundtland Commission). 1987. *Our Common Future*. Oxford University Press, Oxford and New York.

World Resources Institute. 1996. *World Resources: A Guide to the Global Environment 1996–1997*. Oxford University Press, New York.

Promoting peace and security in the oceans

Advisory Committee on Protection of the Sea (ACOPS). 1997. *Oceans and Security: Report of the Conference*, Washington, DC.

Borgese, E.M., ed. 1997. *Peace in the Oceans: Ocean Governance and the Agenda for Peace*, The Proceedings of Pacem in Maribus XXIII, Costa Rica, 1995.

Boutros-Ghali, B. 1995. *An Agenda for Peace* (second edition), United Nations, New York.

Broadus J.M. and Vartanov, R.V., ed. 1994. *The Oceans and Environmental Security: Shared United States and Russian Perspectives*. Island Press, Covelo, California.

Clingan, T.A.. 1980. The next twenty years of naval mobility. *U.S. Naval Institute Proceedings*, 106: 82–93.

Faligot, R., 1997. Le retour des flibustiers, *Politique Internationale*, (Automne), No. 77.

Gold J., 1978. Trust funds in international law: the contribution of the International Monetary Fund to a code of principles, *American Journal of International Law*, 72: 856–866.

ICC International Maritime Bureau. 1997. *Piracy and Armed Robbery Against Ships, Annual Report, 1st January –31st December 1997*. Braking, Essex.

International Institute of Humanitarian Law. 1995. *San Remo Manual in International Law Applicable to Armed Conflicts at Sea*. Cambridge University Press, Cambridge, UK.

International Security Readers. 1988. *Naval Strategy and National Security*, MIT Press, Cambridge, MA.

Larson, D.L. 1993. *Security Issues and the Law of the Sea*. University Press of America, Lanham, Maryland.

Prins, G. and Stamp, R. 1991. *Top Guns and Toxic Whales: The Environment and Global Security*. Earthscan, London.

Pugh, M. 1994. *Maritime Security and Peacekeeping: A Framework for United Nations Operations*. Manchester University Press, Manchester.

Treves, T. 1980. La notion d'utilisation des espaces marins à des fins pacifiques dans le Nouveau Droit de la Mer, *Annuaire Français de Droit International*, 26: 687– 689.

The quest for equity in the oceans

Bedjaoui, M. 1994. *The New World Order and the Security Council*. Martinus Nijhoff, Dordrecht.

Bhagwati, J.N., ed. 1977. *The New International Economic Order: The North–South Debate*, MIT Press, Cambridge, MA.

Cicin-Sain, B. and R. W. Knecht, 1993. Implications of the Earth Summit for Ocean and Coastal Governance, *Ocean Development and International Law*, October–December, pp. 323–353.

European Council on Environmental Law, 1997. *Legal Problems Concerning Bioprospecting for Genetic Resources Located in Marine Hydrothermal Vents Beyond National Jurisdiction.* (Manuscript). Funchal, Madeira, 17 May.

Franck, T. 1995. Equity in international law, in *Perspectives in International Law.* Kluwer Law International, Dordrecht, Boston and London.

United Nations Development Programme, 1997. *Human Development Report 1997.* Oxford University Press, New York and Oxford.

World Bank. 1997. *World Development Report.* Oxford University Press, New York.

Ocean science and technology

Bernstein, P.L. 1996. *Against the Gods: The Remarkable Story of Risk.* John Wiley,

Bridgewater, P. 1997. *Oceans and Biodiversity.* Independent World Commission on the Oceans, Study Group on Science and Technology (SG/ST/WP1), 14–16 March 1997, Lisbon.

Cicin-Sain, B. 1993. Sustainable development and integrated coastal management, *Ocean and Coastal Management* (Special issue Coastal Zone Management), 21: 11–43.

Cook, P.J. 1996. Societal trends and their impact on the coastal zone and adjacent Seas. *British Geological Survey*, Technical Report WQ 96/3.

Earle, S. 1995. *Sea Change: A Message of the Oceans*, G. P. Putnam's Sons, New York.

Gonçalves, M.E. 1983. Science, technology and the new convention on the law of the sea, *Impact of Science on Society*, No. 3–4: 347–354.

Hsu, K.J. and Thiede, J. eds. 1992. *Use and Misuse of the Seafloor.* John Wiley. Chichester.
Pernetta, J.C. and Elder, D.L., 1992. Management and planning for coastal changes. *Ocean and Coastal Management*, 18: 113–160.

Treves, T. 1977. Le transfert de technologie et la conférence sur le droit de la mer, *Journal du Droit International*, 104: 43–65.

Turner, R.K. Subak, S. and Adger, W.N. 1996. Pressures, trends and impacts in coastal zones: interactions between socioeconomic and natural systems. *Environmental Management*, 20(2): 159–173.

Valuing the oceans

Arrow, K., Bolin, B., Costanza, R., Dasgupta, P., Folke, C., Holling, C.S., Jansson, B.-O., Levin, S., Mäler, K.-G., Perrings, C. and Pimentel, D. 1995. Economic growth, carrying capacity, and the environment. *Science*, 268, 28 April: 520–521.

Charles, A. 1994. Towards sustainability: the fishery experience. *Ecological Economics*, 11: 210–211.

Costanza, R.R. d'Arge, R., de Groot, S., Farber, R., Grasso, S., Hannon, M., Naeem, B., Limburg, S. Paruelo, K., O'Neill, J., Raskin, R.V., Sutton, R., and van den Belt, P. M. 1997. The value of the world's ecosystem services and natural capital, *Nature*, London, 15 May.

Dasgupta, P. and Mäler, K-G. 1998. The environment and emerging development issues. *Proceedings of the Annual Conference on Development Economics*, World Bank, Washington, DC, 101–152.

Food and Agriculture Organization of the United Nations, 1997. *Recent Developments in World Fisheries*, FAO, Rome.

Mangel, M., Talbot, L.M., Meffe, G.K., Agardy, T., Alverson, D.L., Barlow, J., Botkin, D.B., Budowski, G., Clark, T., Cooke, J., Crozier, J.H., Dayton, P.K., Elder, D.L., Fowler, C.W., Funtowicz, S., Giske, J., Hofman, R.J., Holt, S.J., Kellert, S.R., Kimball, L.A., Ludwig, D., Magnusson, K., Malayang, B.S., Mann, C., Norse, E.A., Northridge, S.P., Perrin, W.F., Perrings, C., Peterman, R.M., Rabb, G.B., Regier, H.A., Reynolds, J.E., Sherman, K., Sissenwine, M.P., Smith, T.D., Starfield, A., Taylor, R.J., Tillman, M.F., Toft, C., Twiss, J.R., Wilen, J., Young, T.P. 1996. Principles for the conservation of wild living resources. *Ecological Applications*, 6(2): 338–362.

Panayotou, T. 1994. *Economic Instruments for Environmental Management and Sustainable Development*, Harvard Institute for International Development, Cambridge, MA.

Mitchell R. 1994. *International Oil Pollution at Sea: Environmental Policy and Treaty Compliancy*, MIT Press, Cambridge, MA.

Roberts, C. 1997. Ecological advice for the global fisheries crisis. *Trends in Ecology and Evolution*, 12(1): 35–38.

Sandler, T. 1997. *Global Challenges*. Cambridge University Press, Cambridge.

Tolba, M.K., El-Kholy, O.A. El-Hinnawi, E., Holdgate, M.W., McMichael, D.F. and Munn, R.E., eds. 1992. *The World Environment 1972-1992: Two Decades of Challenge*. United Nations Environment Programme, Chapman & Hall, London.

Our oceans: public participation and awareness

Allott, P. 1993. Mare nostrum: a new international law of the sea, in Van Dyke et al., *Freedom of the Seas in the 21 Century: Ocean Governance and Environmental Harmony*. 49–71.

Charnovitz, S. 1997. Two centuries of participation: NGOs and international governance. *Michigan Journal of International Law*, 18: 183–286.

Hewison, G.J. 1996. The role of environmental non-governmental organizations in ocean governance. *Ocean Yearbook*, 12: 32–51.

International Telecommunications Union. 1994. *Multimedia Highways*, Geneva.

March, J.G. and Olsen, J.P. 1995. *Democratic Governance*. Free Press, New York.

Mathews, J.T. 1997. Power Shift. *Foreign Affairs*, 76: 50–66.

Willetts, P. ed. 1996. *The Conscience of the World: The Influence of Non-Governmental Organizations in the United Nations System*, Washington, DC.

Yankelovich, D. 1991. *Coming to Public Judgment: Making Democracy Work in a Complex World.* Syracuse University Press, Syracuse.

Towards effective ocean governance

Chayes, A. and A. 1996. *The New Sovereignty.* Harvard University Press, Cambridge, MA.

Mann Borgese, E. 1995. *Ocean Governance and the United Nations.* Center for Foreign Policy Studies, Dalhousie University, Halifax.

Kimball, L.A. 1997. Whither international institutional arrangements to support ocean law in politics, *Columbia Journal of Transnational Law*, 36: 301–339.

Lévy, J-P. 1991. Une politique marine integrée: objectif réaliste ou illusoire. *Espaces et Resources Maritimes*, Volume V, Pedone, Paris.

Payoyo, P. B., ed. 1994. *Ocean Governance: Sustainable Development of the Seas.* United Nations University Press, Tokyo.

Sand, P.H. 1996. Institution-building to assist compliance with international environmental law: perspectives. *Heidelberg Journal of International Law*, 56: 774.

Schrijver, N. 1997. *Sovereignty over Natural Resources: Balancing Rights and Duties.* Cambridge University Press, Cambridge.

United Nations, 1997. *The Law of the Sea, Official Texts of the UN Convention on the Law of the Sea of 10 December 1982 and of the Agreement Relating to the Implementation of Part XI.* UN Publication, Sales No. E. 97.V.10.

United Nations, 1997. *Renewing the United Nations: A Programme for Reform*, Report by the Secretary-General to the General Assembly, New York.

Van Dyke, J., Zaelke, D., and Hewison, G., eds. 1993. *Freedom for the Seas in the 21st Century: Ocean Governance and Environmental Harmony.* Island Press, Washington, DC.

Annex
D

Acronyms and Abbreviations

A

Agenda 21 (Programme of Action adopted by the 1992 United Nations Conference on Environment and Development)
ASEAN (Association of Southeast Asian Nations)
ASOC (Antarctic and Southern Ocean Coalition of NGOs)

B

Basel Convention (Basel Convention on the Control of Transboundary Movements of Hazardous Wastes and their Disposal)
Brundtland Commission (see **WCED**)

C

CBD (Convention on Biological Diversity)
CD-ROM (compact disk-read only memory)
CEDE (European Council on Environmental Law)
CFCs (chlorofluorocarbon; principal substance responsible for ozone layer depletion)
CFP (Common Fisheries Policy, of the EC)
CITES (Convention on International Trade in Endangered Species of Wild Fauna and Flora)
CMM (Commission for Marine Meteorology, of WMO)
CMS (Convention on the Conservation of Migratory Species of Wild Animals)
CO2 (carbon dioxide)
COFI (Committee on Fisheries, of FAO)
COLREG (Convention on International Regulations for Preventing Collision at Sea, of IMO)
CSD (Commission for Sustainable Development of UN/ECOSOC)

D

DDT (dichloro-diphenyl-trichloro-ethane; an insecticide)
DNA (deoxyribonucleic acid; the molecular basis of heredity)
DOALOS (Division for Ocean Affairs and Law of the Sea, of the UN)

E Annex D

Earthwatch (UNEP assessment programme)
EBRD (European Bank for Reconstruction and
Development)
EC (European Community, or European Union)
ECA (Economic Commission for Africa, of the UN)
ECEL (European Council on Environmental Law)
ECOSOC (Economic and Social Council, of the UN)
EEZ (Exclusive Economic Zone)
EIB (European Investment Bank, of the EC)
EU (European Union, see EC)

F

FAO (Food and Agriculture Organization of the
United Nations)

G

GATT (General Agreement on Tariffs and Trade)
GCC (Cooperation Council for the Arab States of the
Gulf)
GCOS (Global Climate Observing System)
GCRMN (Global Coral Reef Monitoring Network)
GDP (Gross Domestic Product)
GEBCO (General Bathymetric Chart of the Oceans)
GEF (Global Environment Facility)
GEMS (Global Environment Monitoring System, of
UNEP)
GESAMP (Joint Group of Experts on the Scientific
Aspects of Marine Environment Protection)
GFCM (General Fisheries Council for the
Mediterranean, of FAO)
GIPME (Global Investigation of Pollution in the
Marine Environment)
GLOSS (Global Sea Level Observing System)
GNP (Gross National Product)
GOOS (Global Ocean Observing System)
GPA/LBA (Global Programme of Action for the
Protection of the Marine Environment from
Land-Based Activities)
Greenpeace (Greenpeace International Council)
GTOS (Global Terrestrial Observing System)

H

HELCOM (Helsinki Commission, for the Baltic Sea)
HOPE (Hydrothermal Ocean Processes and Systems)

I

IABO (International Association for Biological Oceanography)
IACSD (Inter-Agency Committee for Sustainable Development, of the UN)
IAEA (International Atomic Energy Agency)
IAPSO (International Association for the Physical Sciences of the Oceans)
IATTC (Inter-American Tropical Tuna Commission)
IBRD (International Bank for Reconstruction and Development, or World Bank)
ICAM (Integrated Coastal Area Management)
ICC (International Chamber of Commerce)
ICCAT (International Commission for the Conservation of Atlantic Tuna)
ICES (International Council for the Exploration of the Sea)
ICJ (International Court of Justice)
ICSPRO (Inter-Secretariat Committee on Scientific Programmes Relating to Oceanography)
ICSU (International Council of Scientific Unions)
ICZM (Integrated Coastal Zone Management)
IDA (International Development Association, of the World Bank Group)
IDOE (International Decade of Ocean Exploration)
IFC (International Finance Corporation, of the World Bank Group)
IGOSS (Integrated Global Ocean Services System)
IHO (International Hydrographic Organization)
IHP (International Hydrological Programme)
ILC (International Law Commission of the UN)
ILO (International Labour Organization)
IMA (International Maritime Academy of IMO)
IMCAM (Integrated Marine and Coastal Area Management)
IMF (International Monetary Fund)
IMO (International Maritime Organization)
INCO (Independent Commission on the Oceans, of Brazil)
IOC (Intergovernmental Oceanographic Commission of UNESCO)
IODIE (International Oceanographic Data and Information Exchange)

IOE (Indian Ocean Expedition)
IOI (International Ocean Institute)
IPCC (Intergovernmental Panel on Climate Change)
ISBA (International Seabed Authority)
ITLOS (International Tribunal for the Law of the Sea)
ITQ (individual transferable quota)
IUCN (International Union for Conservation of Nature and Natural Resources, or World Conservation Union)
IWC (International Whaling Commission)
IWCO (Independent World Commission on the Oceans)
IWOF (Independent World Ocean Forum)

L

LDC (London Dumping Convention, Convention on the Prevention of Marine Pollution by Dumping of Wastes and Other Matters)
LOSIC (Law of the Sea Information Circular of DOALAS)

M

MAB (Man and the Biosphere Programme, of UNESCO)
MARPOL (International Convention for the Prevention of Pollution from Ships, of IMO)
MAST (Marine Science and Technology Programme, of the EC)
MCSD (Mediterranean Commission for Sustainable Development)
MEPC (Marine Environment Protection Committee, of IMO)
MOU (Paris Memorandum of Understanding on Port State Control)

N

NATO (North Atlantic Treaty Organization)
NGO (Non-Governmental Organization)

O

ODP (Ocean Drilling Programme)
OG (Ocean Guardian)
OPEC (Organization of Petroleum Exporting Countries)
OSPAR (Oslo/Paris Convention on Protection of the Marine Environment of the North-East Atlantic)
OTEC (Ocean Thermal Energy Conversion)

P

PCB (Polychlorinated Biphenyl)

S

SCAR (Scientific Committee on Antarctic Research, of ICSU)
SCOR (Scientific Committee on Oceanic Research, of ICSU)
SIDS (Small Island Developing States: Programme of Action for the Sustainable Development)
SIPRI (Stockholm International Peace Research Institute)
SPLOS (States Parties to the Law of the Sea Convention)
SPREP (South Pacific Regional Environment Programme)

T

TED (Turtle Excluder Device, in fishing gears)
TEMA (Training, Education, Mutual Assistance and Capacity Building)
TNC (Transnational Corporation)
TOGA (Tropical Oceans and Global Atmosphere Programme)

U

UN (United Nations)
UNACC (United Nations Administrative Committee on Coordination)
UNCED (United Nations Conference on Environment and Development)
UNCLOS (United Nations Convention on the Law of the Sea)
UNCTAD (United Nations Conference on Trade and Development)
UNDP (United Nations Development Programme)
UNEP (United Nations Environment Programme)
UNESCO (United Nations Educational, Scientific and Cultural Organization)
UNFCCC (United Nations Framework Convention on Climate Change)
UNGA (United Nations General Assembly)
UNIDO (United Nations Industrial Development Organization)
USD (United States Dollar)

WBCSD (World Business Council on Sustainable
Development)
WCED (World Commission on Environment and
Development, or Brundtland Commission)
WHO (World Health Organization)
WMO (World Meteorological Organization)
WMU (World Maritime University, of IMO)
WOAO (World Ocean Affairs Observatory)
WOCE (World Ocean Circulation Experiment)
World Bank (see **IBRD**)
World Conservation Union (see **IUCN**)
WTO (World Trade Organization)
WWF (World Wide Fund for Nature)
WWW (World Weather Watch, of WMO)

Annex E

The Commission and its Work

The Independent World Commission on the Oceans was launched in December 1995, in Tokyo, in recognition of the fundamental importance of the oceans in planetary survival, in the maintenance of peace and security as well as in the development of human society.

The Commission shares with a number Independent Commissions which have preceded it – especially Willy Brandt's Commission on International Development Issues, Olaf Palme's Commission on Disarmament and Security Issues, Gro Harlem Brundtland's World Commission on Environment and Development and, most recently, Ingvar Carlsson's and Shridath Ramphal's Commission on Global Governance – a concern for the dignity and equality of human beings in both present and future generations.

While continuing in this grand tradition, the Commission has had to face its own challenges. Humankind has not yet succeeded in adapting its perceptions and re-adjusting its institutions to the growing reality that the oceans should no longer be considered an unlimited source of wealth, opportunity and abundance. The oceans now impose limits on human activity at the same time as they are revealing new potentials and opportunities.

The Commission has responded to the intensified competition between different uses of the oceans, as it has sought ways to overcome the political, economic and behavioural obstacles to the development of integrated mechanisms for managing marine activities in a peaceful, equitable and sustainable manner. The Commission believes that the needed public awareness, of the vital importance of the oceans to humankind, should be enhanced by giving civil society the opportunity to participate in, or influence, decisions on ocean affairs.

THE COMMISSIONERS

Chairman

* Mário Soares, Portugal
President of the European Movement. President of the Portugal Africa Foundation. Former President of the Republic of Portugal; former Prime Minister; former Minister of Foreign Affairs. Former Professor. Practised law for many years.

Vice-Chairmen

* Abdulmohsen Al-Sudeary, Saudi Arabia
Former Ambassador and former President, International Fund for Agricultural Development (IFAD), Rome. Member of the Board of Directors of the *Environment & Development Magazine*.

* Kader Asmal, South Africa
Minister of Water Affairs and Forestry and Member of Parliament. Chairman of the World Commission on Dams, launched in November 1997 by IUCN and the World Bank. Former Professor of Human Rights at the University of the Western Cape.

* Elisabeth Mann Borgese, Canada (until 24 February 1998)
Professor of Political Science and Professor of Law, Dalhousie University. Founder and Honorary Chair, International Ocean Institute (IOI), headquartered in Malta. Co-editor of the Ocean Yearbook, University of Chicago Press.

* Eduardo Faleiro, India
Former Minister of State for Chemicals, Fertilizers and Ocean Development. Former Minister of State for External Relations. Former Minister of State for Finance. Advocate of the Supreme Court of India.

* Patrick Kennedy, USA
Member, House of Representatives for Rhode Island. Member of, e.g. the House National Security Committee and the Resources Committee (and serves on the Subcommittee on Fisheries, Wildlife and Oceans). Co-author of the National Oceanographic Partnership Act.

* Ruud Lubbers, Netherlands
Minister of State. Former Prime Minister (1982–1994) and Minister for Economic Affairs (1973–1977). Professor of Globalization, Center for Economic Research, Catholic University of Brabant. Visiting lecturer, John F. Kennedy School of Government, Harvard University.

* Guido de Marco, Malta
Former Deputy Prime Minister and Minister Foreign Affairs. Former President, United Nations General Assembly. Professor of Criminal Law, University of Malta.

* Yoshio Suzuki, Japan
Member, House of Representatives. Shadow Minister for Economy & Finance of the New Frontier Party. Former Chief Counsellor, Nomura Research Institute. Former Executive Director of the Bank of Japan and Director of the Institute for Monetary and Economic Studies.

* José Israel Vargas, Brazil
Minister of Science and Technology. President, Third World Academy of Sciences. Former Vice-Chairman and Chairman, Executive Board of UNESCO. Former Chairman, Committee for Science and Technology, ILO.

Members

Seyyid Abdulai, Nigeria
Director-General, OPEC Fund for International Development, Vienna. Former Executive Director, World Bank. Former Managing Director, Federal Mortgage Bank of Nigeria.

Najeeb Al-Nauimi, Qatar
Former Minister of Justice. Agent and Counsel to the International Court of Justice in the case concerning Maritime Delimitation and Territorial Questions (Qatar vs. Bahrain Case). Professor of Public International Law at Qatar University.

Oscar Arias, Costa Rica
Nobel Laureate for Peace (1987). Former President of the Republic, former Minister of Planning and Economic Policy, and former Professor of Political Science at the University of Costa Rica. Founder of the Arias Foundation for Peace and Human Progress.

Alicia Bárcena, Mexico
Senior Advisor, Global Environmental Citizenship Programme, UNEP. Executive Director, Earth Council (1992–1995). Principal Officer on Oceans, Coastal Development and Living Marine Resources (1990–1992), UNCED. Under-Secretary of Ecology, Government of Mexico (1982–1986).

Mohammed Bedjaoui, Algeria
Judge and Former President, International Court of Justice. Former Secretary General of the Government of Algeria; Ambassador to France (1970–1979) and to the United Nations (1979–1982); Head of Delegation to the UN Conference on the Law of the Sea (1976–1980).

Driss Ben Sari, Morocco
Professor of Geophysics. Elected Member of the General Committee of ICSU (1993–   ). Former Director, Moroccan National Centre for Coordination and Planning of Scientific and Technical Research (1979–1994).

Patricio Bernal, Chile
Oceanographer. Executive Secretary, Intergovernmental Oceanographic Commission (IOC/UNESCO), Paris (1998–   ). Former, Technical Adviser, National Commission of the Environment. Former Under-Secretary of State for Fisheries.

Peter Bridgewater, Australia
Chief Science Adviser Environment Australia. Former Chairman, International Whaling Commission (1995–1997). Chair, Intergovernmental Co-ordinating Council Man and the Biosphere Programme (UNESCO). Former Chief Executive Officer, Australian Nature Conservation Agency.

Ian Burton, Canada
Independent scholar and consultant. Former Senior Policy Advisor, Environment Canada; Former Director Environmental Adaptation Research, Atmospheric Environment Service (Canada); Former Director, International Federation of Institutes for Advanced Study.

* Salvino Busuttil, Malta (Treasurer)
Executive President, Malta Association of Human Rights; Vice-President, Mediterranean Centre for Human Rights. Former Director-General, Foundation for International Studies, University of Malta. Former Director, UN Mediterranean Action Plan.

Lucius Caflisch, Switzerland
Judge, European Court of Human Rights. Professor of International Law, Graduate Institute of International Studies, Geneva. Legal Advisor, Federal Department of Foreign Affairs, Bern. Deputy Head of Delegation to the UN Conference on the Law of the Sea (1974–1982).

Ricardo Diez-Hochleitner, Spain
President of The Club of Rome. Former Under-Secretary of State for Education and Science (1969–1972). Permanent member, Spanish National Research Council. President, International Advisory Board of EXPO 2000.

† René-Jean Dupuy, France
Professor of International Law, Collège de France, Membre de l'Institut. President of the Institute of International Law. Former Secretary General of the Hague Academy of International Law.

Richard Falk, USA
Professor of International Law and Practice, Centre of International Studies, Princeton University. Editorial Board, American Journal of International Law. Board of Directors, Institute for Defence and Disarmament Studies.

B. A. Hamzah, Malaysia
Director-General, Malaysian Institute of Maritime Affairs (MIMA). Diploma programme in the Law of the Sea, Institute of Social Studies, The Hague. Former assistant Director-General, Senior Fellow, and Fellow, Institute of Strategic and International Studies (ISIS).

Klaus-Jürgen Hedrich, Germany
Parliamentary State Secretary, Federal Ministry for Economic Cooperation and Development. Member, Parliamentary Committee on Economic Cooperation (1983–1994). Chairman, German ASEAN Parliamentarians Association (1987–1994).

* Sidney Holt, UK (General Editor) (until November 1997)
Specialist, Management Marine Living Resources, formerly with FAO, UNESCO, IOC and UNEP. Fisheries research in the UK. Advisor to several non-governmental environment organisations and to some governments.

† Danielle Jorre de St Jorre, Seychelles
Minister for Foreign Affairs, Planning and Environment. Governor for Seychelles on the Board of Governors for the World Bank and African Development Bank. Vice-President (African Region), Advisory Committee on the Protection of the Sea (ACOPS).

Stjepan Keckes, Croatia
Former Director, Regional Seas and Oceans and Coastal Areas Programmes (UNEP, 1975–1990). Member, Group of Expert on Scientific Aspects of Marine Environmental Protection (GESAMP); Co-Chairman, GESAMP working Group for Marine Environmental Assessments.

† John Kendrew, UK
Nobel Chemistry Laureate. Former Chairman, International Council of Scientific Unions (ICSU). Former President of St. John's College, Oxford. Former Director-General of the European Molecular Biology Laboratory, Heidelberg.

Tommy Koh, Singapore
Ambassador-at-Large. Director, Institute of Policy Studies. Special Advisor to the Administrator of the UN Development Programme. Chairman, Preparatory Committee and Main Committee, UNCED (1990–1992). President, Third UN Conference on the Law of the Sea (1981–1982).

Nicolay P. Laverov, Russia
Vice-President, Russian Academy of Sciences. Director, Institute of Ore Deposits Geology, Russian Academy of Sciences. Chairman, Scientific Council for Realization of the State Scientific and Technical Programme 'Global Nature and Climate Changes'. Former President, Kirghiz Academy of Sciences.

Ulf Lie, Norway
Professor, Centre for Studies of Environment and Resources, University of Bergen, Former Chairman, Intergovernmental Oceanographic Commission (IOC). Former Research Associate Professor, University of Washington. Former Programme Specialist, UNESCO.

Luiz Filipe de Macedo Soares, Brazil
Ambassador to India, Maldives, Nepal and Sri Lanka. Head of Delegation to the: Preparatory Commission for the Seabed Authority; Climate Change Convention Negotiations; and Preparatory Commission for UNCED.

Donald Mills, Jamaica
Ambassador. Chairman, Commonwealth Foundation. Former Alternate Executive Director on the Board of the International Monetary Fund; Ambassador to the UN; Chairman of the Group of 77; President of the Economic and Social Council (ECOSOC); President of the UN Security Council.

Venâncio de Moura, Angola
Minister of Foreign Affairs. Former First Ambassador of Angola; First Deputy-Minister of Foreign Affairs; Negotiator on the Angola peace process; and Minister of External Affairs and Cooperation.

Noriyuki Nasu, Japan
Professor (1984–1994), University of the Air. Professor (1962–1984) and Director (1968–1972, 1980–1984), Ocean Research Institute, and Professor (1977–1979), Earthquake Research Institute, University of Tokyo. Advisor to the Government of Japan, Prefectural Governments, and companies.

Alassane Dialy Ndiaye, Senegal
Minister of Fisheries and Shipping. Former Director-General, Senegalese Telecom; President of the Meeting of Signatories of INTELSAT ; Promoter of the Submarine Cables Senegal/Brazil and Senegal/Portugal.

Carlo Ripa di Meana, Italy
Member of the European Parliament. Former Commissioner for the Environment of the European Community. Former Minister of the Environment.

* Mário Ruivo, Portugal (Coordinator)
Professor of Ocean Policy and Management, University of Oporto/ICBAS. Chairman, National Council for Environment and Sustainable Development. Former Minister of Foreign Affairs. Former Secretary, Intergovernmental Oceanographic Commission (IOC/UNESCO), Paris.

Esekia Solofa, Western Samoa
Vice Chancellor, University of the South Pacific (USP) Director, Institute of Social Administrative Studies, and Director, Planning and Development, USP (1983–1991). CEO, Public Service Commission, Western Samoan Government (1981–1982).

Jilan Su, China
Director and Chief Scientist, Second Institute of Oceanography, State Oceanic Administration; Vice-Chairman, Standing Committee Earth Sciences Division, Chinese Academy of Sciences; Vice-Chairman, Standing Committee, Chinese Society of Oceanography.

Alexander Yankov, Bulgaria
Judge, International Tribunal for the Law of the Sea, Hamburg. Professor of International Law, Bulgarian Academy of Sciences. Member of the International Law Commission. President, Bulgarian Association of International Law. Former Ambassador to the UK and the UN.

† Deceased * Member of the Executive Committee

The Commission wishes to express its high appreciation to former members **Elisabeth Mann Borgese** and **Sidney Holt** for their valuable contributions to its work.

The Commission greatly benefited from the wisdom of the following 'eminent persons': **Sylvia A. Earle, Heitor Gurgulino de Souza, Carl-August Fleischhauer, Tsutomu Fuse, Ronald St. J. Macdonald, Pierre Papon, Claiborne Pell, António Ruberti, Juan Somavia and Yevgueni Velikov.**

During its mandate, ending in 1998, the Commission will seek to develop world consciousness of the unique role of the oceans, including their interaction with rivers and land-based activities, for planetary survival, and the critical importance of rational management of ocean space and resources. In pursuit of these aims, the Commission will:

● draw the attention of world leaders (including political, business, environmental, scientific and educational leaders), of NGOs and the public at large (men and women and particularly the young) to relevant issues of ocean development and the direct or indirect impact of human activity on ocean resources;

● encourage the further development of the ocean regime evolving from the United Nations Convention on the Law of the Sea in the light of changing scientific perceptions and discoveries, with particular attention to the needs of developing countries;

● study the interaction between the United Nations Convention on the Law of the Sea and other related legal instruments and programmes of action (in particular Agenda 21 of UNCED), and explore ways to promote their implementation, taking account of overlaps, complementarities and synergies;

● examine the existing economic potential of the oceans, including the proper use of their living and non-living resources, and explore non-conventional uses of ocean resources and services emerging from new developments in science and technology;

● promote the incorporation of the marine dimension in national developments plans;

●analyse the requirements of integrated coastal zone management in the light of the pressures generated, inter alia, by the growth of the population, tourism and trade, and taking account of the conclusions and recommendations of relevant international conferences (UNCED, Population, Habitat, Small Island Developing States);

● explore new forms of North–South and South–South cooperation for joint technology development;

•study the threats to the seas and oceans and the sustainability of their resources and various uses, including the potential social and economic impact of global warming and sea-level rise;

• endeavour to define ways of strengthening the institutional framework for ocean governance at various levels;

• contribute to the development of the peaceful uses of the oceans and examine the potential of ocean governance for the implementation of the United Nations Secretary-General's Agenda for Peace.

In the fulfilment of its tasks, the Commission will encourage the ratification and implementation of the UN Convention on the Law of the Sea as well as the implementation of Agenda 21 of UNCED (in particular, Chapter 17).

Furthermore, the Commission will cooperate closely with the United Nations, UNESCO/IOC and other agencies and programmes of the United Nations system competent in ocean affairs, and with other inter-governmental organizations as well as with non-governmental organisations, at the national, regional, and global levels.

WORKING FOR THE COMMISSION

The Commission established its secretariat in Geneva – which was as from May 1996 located at 14 Avenue de Joli-Mont – with the support of the governments of the Canton of Geneva and the Swiss Federation. Its staff members were:

Secretary General (until 5 March 1998)
Layashi Yaker *

Executive Secretary
Jean-Pierre Lévy *

* ex officio member of the Executive Committee

Professional Staff
Jan van Ettinger   Senior Officer (Management)
Thomas Ganiatsos   Senior Officer (Substance)

General Support Staff
Lydia Beauquis   Senior Secretary
Sheahan Fernando   Administrative Assistant
Fatima Matallana   Secretary

As to the Commission's financial administration, the secretariat first benefited from the relevant experience and friendly advice of Tudor Jayawardana, Administrator of the South Centre, and, starting 1997, from the professional assistance, as part-time Administrator, of Louisa Seiringer. Her financial statements have been audited by Henry Ferguson and David Curry of KPMG S.A., Geneva office.

As to the media and editing and publishing of the Commission's Report, the secretariat enjoyed the valuable assistance, in alphabetic order, of: Peter Cook, Antony Dolman, Sylvia Earle, Richard Falk, Sidney Holt, Stjepan Keckes, Lee Kimball, John May, Charles Perrings, Chitra Radhakishun, Paul Ress, and Peter Sand.

The Commission's Chairman had the following staff support in his Lisbon office:

Mário Baptista Coelho   Personal Assistant to the Chairman

Ana-Maria Casquilho Plimer   Personal Assistant to the Coordinator

Ana-Teresa Egea   Secretary
Rita Pargana   Secretary

During its three-year period (1995–1998), the Commission held six Plenary Sessions:

First Session
Tokyo (Japan), 13–15 December 1995

The launching Session took place at the United Nations University (UNU). At its opening, welcoming addresses were delivered by the Rector of the UNU and the Governor of Tokyo and messages from the Government of Japan were read. During the Session, among others, the Purpose and Terms of Reference were adopted and a Programme of Work was considered.

Second Session
Rio de Janeiro (Brazil), 2–5 July 1996

At the opening, a message was read from the President of Brazil, Fernando Henrique Cardoso. During the Session, the Commission considered the four key issues already selected (legal and institutional framework; peaceful uses; ocean economics; and science and technology), and selected two other issues (public awareness; and partnership and solidarity).

Third Session
Rotterdam (Netherlands), 26–29 November 1996

The opening was attended by Her Majesty Queen Beatrix and was addressed by the Minister of Transport, Public Works and Water Management and on behalf of the Minister for Development Cooperation. During the Session, among others, the report of the Study Group on Legal and Institutional Issues was discussed, and the organization of five other Study Groups was considered.

Fourth Session
Rhode Island (USA), 6–9 June 1997

At the opening, the Under-Secretary of State for Global Affairs, Timothy Wirth, assured that the ratification of UNCLOS was a top priority for the present US Administration. A special presentation on US Marine Science and Technology took place. During the Session, among others, the Commission considered the reports of its six Study Groups and discussed the outline of its Final Report.

Fifth Session
Cape Town (South Africa), 11–14 November 1997

At the opening, in his key address, President Nelson Mandela advocated, among others, to base ocean policy on the primacy of people and their long-term well-being. During the Session, the Commission discussed especially the first draft of its Final Report, in particular its message and recommendations. To prepare this discussion, two sessional drafting groups were established.

Extra Session
Rabat (Morocco), 6–8 February 1998

At the opening, HRH Prince Sidi Mohamed, heir-apparent to the throne, presented a message of welcome from HM King Hassan II. Upon the invitation of HM the King and co-hosted by the Royal Academy of Morocco through Dr Abdellatif Berbich, this Session completed the discussion of the Report, authorized the Chairman to submit for publication a final draft reflecting this discussion, and reviewed a draft of a Lisbon declaration.

Sixth and Final Session
Lisbon (Portugal), 30 August to 1 September 1998

The Final Plenary Session was organized in Lisbon within the framework of the International Year of the Oceans, 1998, and in conjunction with EXPO'98 'The Oceans: A Heritage for the Future'. It consisted of the public presentation, in particular to the young people on 31 August 1998, of the Commission's Final Report and a high-level segment, during which the Lisbon Declaration was adopted.

Meetings of the Executive Committee took place immediately before the Plenary Sessions of the Commission, and have been convened intersessionally in Lisbon (Portugal) and in Geneva (Switzerland).

STUDY GROUPS

Drawing upon available information from a wide variety of sources, the Commission has carried out an integrated, action-orientated review of ocean affairs. This review covered six key issues that have been explored in depth. For each of them, the Commission convened a Study Group involving a number of its members as well as outside experts, some of whom were invited to prepare background papers. These six Study Groups were:

● Peaceful uses of the ocean, security and sovereignty

Chairman:	Guido de Marco
Rapporteur:	Richard Falk
Outside experts:	Geoffrey Lynn Chesbrough
	Yves Leenhart
	David MacTaggart

● Legal and institutional framework for use and protection of the ocean

Chairman:	Luiz Filipe de Macedo Soares
Rapporteur:	Lee A. Kimball
Outside experts:	Patricia Birnie
	Carl-August Fleischhauer
	Ahmed Mahiou
	Thomas Mensah

● Economic uses of the ocean in the context of sustainability

Chairman:	Ruud Lubbers
Rapporteur:	Paul Streeten
Outside experts:	Anthony Charles
	Robert Costanza
	Catrinus Jepma
	Charles Perrings

● Marine science and its technology

Chairman :	Ulf Lie
Rapporteur:	William Andahazy and
	Peter Cook
Outside experts:	Biliana Cicin-Sain
	Ehrlich Desa
	Sylvia Earle
	Kees Lankester
	Pierre Papon

● Partnership and solidarity : North/South issues

Chairman:	Kader Asmal
Rapporteur:	Nazli Choucri
Outside experts:	Havelock Brewster
	Gamani Corea
	Gotfried Hempel
	José Luis de Jesus
	Gunnar Kullenberg
	Timothy Shaw
	Manuel Tello

● Public awareness and participation

Chairman:	Sidney Holt
Rapporteurs:	Sidney Holt and Peter H. Sand
Outside experts:	Anil Agarwal
	Margrete Auken
	Michael Donoghue
	Naoko Funahashi
	Maria Eduarda Gonçalves
	Mathias Kaiser
	John May
	Lesley Sutty

PAPERS PREPARED FOR THE COMMISSION

- William Andahazy and Peter Cook – Ocean Science and its Technology
- Scott Barrett – Economic Incentives and the Oceans
- Anthony T. Charles – Sustainable Coastal Fisheries
- Nazli Choucri – Partnership and Solidarity: North/South Issues
- Robert Costanza – The Ecological, Economic, and Social Importance of the Oceans
- Sylvia A. Earle – An Overview of the Future of the Ocean
- Hilal Elver – Aegaen Sea Dispute between Greece and Turkey
- Richard Falk – On the Peaceful Uses of the Ocean: Sovereignty, Security, Sustainability and Development
- Zhiguo Gao – Maritime Claims and Disputed Islands in the South and the East China Seas: Options and Approaches
- Sidney Holt and Peter H. Sand – Our Ocean: Public Awareness and Participation
- Katia Katsigera – The Greco-Turkish Dispute over the Aegean Continental Shelf
- Stjepan Keckes – Review of International programmes relevant to the work of the Independent World Commission on the Oceans
- Lee A. Kimball – Legal and Institutional Framework for Use and Protection of the Ocean
- Kaoru Okuizumi – The East China Sea Dispute
- Charles A. Perrings – The Economics of Ocean Resources
- Charles A. Perrings – The Economic Uses of the Oceans for Sustainable Development
- Peter H. Sand – Ocean Governance: Legal and Institutional Framework
- Paul Streeten – Economic Uses of the Oceans for Sustainable Development

CONTRIBUTIONS

To cover the 1995–1998 expenditures, cash and in-kind contributions have been secured from different sources in Brazil (Government and various public and private contributions), Canada (private contribution), Germany (Government), India (Government), Japan (Nippon Foundation), Morocco (Government), the Netherlands (Government), Portugal (Government), Saudi Arabia (HRH Prince Sultan Ben Abdel-Aziz Al Saud and HRH Prince Talal Bin Abdul-Aziz Al Saud), South Africa (Government and various public and private contributions), Switzerland (Federal Government and Canton of Geneva) and USA (various private contributions) as well as from the Commission of the European Union, OPEC Fund for International Development, UNDP/UNOPS and UNESCO/IOC.

FOLLOW-UP

A group of 'friends' of the IWCO has been formed consisting, in October 1997, of the Permanent Representatives to the UN (New York) of Angola, Brazil, Chile, Fiji, India, Italy, Jamaica, Japan, Malta, Mexico, Morocco, the Netherlands, Portugal, Saudi Arabia and South Africa. Their information note of 10 October 1997 to the UN Secretary-General on the Commission was circulated as an official document of the General Assembly (A/52/458) on 14 October 1997 under agenda item 39 (Oceans and Law of the Sea). As is hoped, this group will be joined by Permanent Representatives of other countries and will submit an Information Note on the work of the Commission, based on its Final Report, to the 53rd General Assembly in Autumn 1998, when it reviews the International Year of the Ocean.

ACKNOWLEDGMENTS

In addition to those mentioned above, the Commission received the support of numerous other persons and organizations. Among these the collaboration with, and support of, EXPO'98, and the Mário Soares Foundation should be acknowledged. Last but not least, the 'local organizers' of the Plenary Sessions of the Commission deserve to be mentioned:

- Tokyo (Japan): Ms Masako Otsuka
- Rio de Janeiro (Brazil): Ms Marcia Graça Melo
- Rotterdam (the Netherlands): Ms Marjolein Matthijssen
- Newport, Rhode Island (USA): Mr Marty Alford
- Cape Town (South Africa): Mr John Cooper
- Rabat (Morocco): Mr Madani
- Lisbon (Portugal): Ms Ana-Maria Casquilho Plimer

Index

The Index was compiled by:
Mary Orchard,
INDEXING SPECIALISTS
of Hove (UK)

Produced for
the
Independent World Commission
on the Oceans
by
John May
Design and Illustration:
Andy Gammon
in
Lewes (UK)

Research for Annex A:
Kieran Mulvaney
Connie Murtagh
Bruce McKay
in
Washington D.C. and Montreal

Thanks are due to the following persons
at Cambridge University Press:
Simon Mitton
Matt Lloyd
Miranda Fyfe
Mary Sanders
Stephanie Thelwell

May 1998

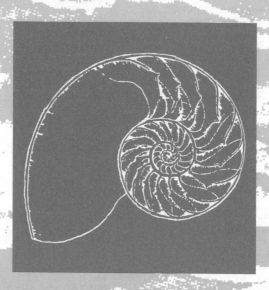

248